Frontiers 01

Frontiers01

Science and Technology, 2001–02

Edited by **Tim Radford** Foreword by **Sir John Sulston**

Atlantic Books
London

First published in 2002 by Atlantic Books,
on behalf of Guardian Newspapers Ltd.
Atlantic Books is an imprint of Grove Atlantic Ltd.

10 9 8 7 6 5 4 3 2 1

A CIP catalogue record for this book is available
from the British Library

ISBN 1 903809 23 1

Printed in Great Britain by St Edmundsbury Press
Design by Ghost

Grove Atlantic Ltd
29 Adam & Eve Mews
London W8 6UG

Contents

Notes on contributors

Mike Baillie is Professor of Palaeoecology at Queens University, Belfast. He is the author of *Exodus to Arthur: Catastrophic Encounters with Comets.*

Sarah Boseley is health editor of the *Guardian*.

Brian Butterworth is Professor of Cognitive Neuropsychology at University College, London, and the author of *The Mathematical Brain*.

Frank Close is a professor of physics at Oxford University. He is also Professor of Astronomy at Gresham College, London and author of *Lucifer's Legacy*.

Andrew Coates heads the Space Plasma Physics Group at the Mullard Space Science Laboratory, University College, London.

Henry Gee is a senior editor with *Nature*. He is the author of *Before the Backbone* and *Deep Time*.

Alastair Hay is a toxicology professor in the Department of Chemical Pathology at Leeds University.

Joe Herbert is Professor of Neuroscience at Cambridge.

Tim Hubbard is head of the Human Genome Analysis Group at the Wellcome Trust Sanger Institute, Cambridge.

David E. H. Jones writes the 'Daedalus' column for *Nature*.

Paul Murdin is a senior fellow at the Institute of Astronomy, Cambridge. He was formerly Director of Science at the British National Space Centre.

Tim Radford is science editor of the *Guardian*.

Sir Martin Rees is Astronomer Royal. He is the author of *Just Six Numbers* and *Our Cosmic Habitat*.

Duncan Steel is a physicist at the University of Salford. His latest book is *Target Earth*.

Mike Stratton leads the Cancer Genome Project at the Wellcome Trust Sanger Institute in Cambridge, and is also at the Institute of Cancer Research. He was responsible for discovering BRCA2, one of the two genes known to predispose women to breast cancer.

Sir John Sulston, founder and until recently Director of the Wellcome Trust Sanger Institute in Cambridge, is a leading figure in the international partnership that is sequencing the human genome.

Ben Wisner is vice-chair of the International Geographical Union's Commission on Hazards and Risk, and a visiting research fellow at the Institute of Development Studies at the London School of Economics.

Preface

Tim Radford

Two British scientists won the Nobel prize for physiology and medicine in 2001. They were Sir Paul Nurse, director of the Imperial Cancer Research Fund, and his colleague Tim Hunt. They shared it with an American, Leland Hartwell, a cancer researcher in Seattle, and all three were awarded science's greatest prize for a series of interconnecting discoveries in the machinery of life – the steps that make a single cell divide. All life starts with a single cell. Human life ripens with an adult made up of around 100 million million cells. Cell division brings life, and when the cell becomes cancerous, it begins to take life away again. So, knowing why cells divide, and why they divide when they do, and why they then stop dividing or, even worse, do not stop, would have profound consequences. Nevertheless, the telephone call from Stockholm was a surprise. Sir Paul told reporters: 'I started running around like a headless chicken until I managed to get a proper message. My office calmed me down when I managed to phone them. When I spoke to Tim, he didn't believe it, he wanted to see it in electronic ink.'

The Nobel award for two much-liked and widely admired men was a high point for British scientists in 2001, but it was an award for science achieved decades ago. Biology, precisely because of such discoveries, has moved on. This book is about some of the landmarks and horizons of science in 2001. It is about dramatic, first-past-the-tape events such as the touchdown of a spacecraft on the surface of a faraway asteroid; or the completion of the entire DNA code of *Yersinia pestis*, the little bacillus behind the Black Death of the Middle Ages; or the publication of the first great wave of research papers on the 3 billion letter code of the human genome itself, officially 'completed' in 2000, but in fact still being tidied up in laboratories in Britain, Europe, America and Japan. This is also a book about decisions: to end the eventful career of Mir, the Russian space station, and carry on with the costly International Space Station, already one of the brightest lights in the night sky; or the controversial decision in Britain to go ahead with embryo stem cell research in the hope of finding new treatments for so-far intractable illnesses.

Some of the essays in this book are an exercise in map-reading: where are we now, in our understanding of the universe, of memory, or of the process of evolution? For the movie-goers of the sixties, the year 2001 was a landmark because of the title of science fiction's most extraordinary and famous film: *2001: A Space Odyssey*. And one of the essays here takes another look at Stanley Kubrick and Arthur C. Clarke. But all science is unfinished business and the stories told in this book can only hint at the dizzying diversity of unfinished business in 2001. It was also the year that one British team showed that one-year-old babies could 'remember' music they heard in the womb, while another team produced evidence that the better the professional footballer, the longer his ring finger was compared to his index finger. It was the year an American astronomer-historian used the position of the planet Venus on a canvas to calculate the date and time that Vincent Van Gogh painted *White House At Night* (it was at 8pm on 16 June 1890) and it was also the year an Italian geophysicist blamed both Nessie the Loch Ness Monster and the Oracle at Delphi on subterranean rumblings along active geological faults. Nessie and Vincent Van Gogh, babies' memories and the development of athletes are perennial puzzles: more of such things, perhaps, in the next edition of *Frontiers*.

This book is indebted to scientists everywhere who produced lively ideas in 2001, and particularly to those who set aside demanding research time to contribute an essay. It could not have been begun without the encouragement of Alan Rusbridger, editor of the *Guardian*. It could not have been completed without the support of Mathew Clayton of the *Guardian* and Toby Mundy and Alice Hunt of Atlantic Books. And it could not have happened at all without public support for the great adventure of science itself. On the day Paul Nurse won the Nobel prize, I reminded him of an irony: nine years earlier he had predicted that British scientists would not be winning many Nobel prizes in the future because of the then government's attitude to pure research. It is worth repeating what he said, because although both government and attitudes have changed, the warning remains as true as ever. 'They thought it could be run by big business and they pushed it hard in that direction, in ways that were counterproductive,' Sir Paul said. 'Indeed, we have to carry out research in the real world and that is absolutely right, but you don't do it, in my view, by not paying people enough, by letting the laboratories decay, by turning the universities into a disgrace.'

Foreword: why we do science

Sir John Sulston

2001 saw the start of 'Science Year' – a series of festivals and events taking place around the country to celebrate the excitement and achievements of science. Wonderfully varied in style and insight, the essays collected here are a fine contribution to that celebration. They are about the search for understanding, the joy of discovery and the curiosity that ensures that the answer to every question leads directly to more questions. This is terrific because science is essentially a cultural activity. It generates pure knowledge about ourselves and about the universe we live in, knowledge that continually reshapes our thinking. While it cannot provide answers to the big questions about the meaning of our lives, it certainly provides an ever-better framework in which to pose them.

But scientific discoveries, in themselves benign, can be used for good or ill, subject to commercial and social imperatives. For example, in recent years it has been biology's turn to lose its innocence. We are living through its Klondike period – the gene rush. Soon the bubble will burst, but for the moment, there is a land grab. There are huge queues of patent applications for DNA sequences with nominal utility. Gene collectors travel the globe to plunder the heritage of local communities for specimens, and then try to sell them back with nominal improvements to their original users.

All this frenzied activity results in some useful products, but it comes at a price. The assertion of ownership over information erodes the free exchange of knowledge, which is an essential part of the wellbeing of science. The full value of a scientific discovery is seldom apparent straight away. So the most effective way forward is to keep as much knowledge as possible in the public domain, where everyone can think about it and use it for further advances.

In human terms, the most important applications of scientific discovery may not be those that generate large revenues, but rather those that take place in universities and research institutes in the poorer parts of the world. The more that is done outside the rich nations the better, for this builds new skills and education where they are lacking, and therefore helps to reduce the wealth gap. The future prosperity of mankind, not to mention its security, lies with developing a more equitable version of globalization. There are encouraging signs that this view is prevailing and that we shall continue to enjoy ever-increasing knowledge and understanding on our voyage of discovery. Have fun with the essays!

Life: can you figure it out?

Tim Hubbard

The epic decipherment of the human genetic code turned out to be just the beginning of an even greater adventure, involving new sciences called proteomics and bioinformatics

A human being is just another machine. This is not an argument about the mysteries of consciousness; just about the machine you use to carry your mind about in. It is an uninterrupted mobile power supply and a source of physical information, but that is about all. Given that most things that go wrong with us are due to physical breakdowns of this machine, it would be nice to understand it a bit more, since that would almost certainly lead to a greater understanding of why these breakdowns occur. For all the marvels of medicine, when you visit your family doctor and say that you don't feel very well, the questions and tests now available don't really provide the kind of detail necessary to know exactly what has happened.

Compare this to taking your car to the garage. For each car there is a complete list of parts and a manual explaining precisely which part goes where, what it does and which other parts it interacts with – and you would be pretty upset if your garage claimed it couldn't fix your car because they had no idea what parts did. Of course, the human body is a bit more complicated than a car. A car is made up of hundreds of parts. Some parts are present only once (exhaust, aerial); others are present in a few copies (four wheels, 16 valves, etc). At the physical level that we can all see, there are 'parts' of the body that appear to have clear functions – heart, arms, eyes etc. Inside the human body, however, these 'parts' are in turn made up of smaller components. The body is made up of roughly 100 million million cells. Human cells come in more than 200 types of widely different abundance. The visible 'parts' of a human body are each a cocktail of different cell types, organized in complex ways.

Yet not even the cell is a small enough unit to consider, if we wish to understand the detail of the human machine. Each cell is a complete machine in its own right and is made up of billions of molecules. Do we have a full parts list at

The genome really is all of it – a complete set of instructions describing how to make a human... If printed on A4 paper, it would cover 750,000 pages

this level? No, but we do know one thing – each cell manufactures most of its components on site. We do not have the list of what gets made, but we do for the first time have the complete list of instructions to tell the cell what to make and how to make it – the human genome sequence.

The acclaim that greeted the completion of the first complete draft of the human genome in June 2000, and the publication of the first papers in February 2001, was justified. The genome really is all of it – a complete set of instructions describing how to make a human. Although the sequencing of the genome is still not finished (as of today 50 per cent is letter perfect and 50 per cent is still draft, with gaps every few tens of thousands of letters), to have got to this point years earlier than expected is a huge technical achievement. The genome is an instruction book that is 3 billion letters long. If printed on A4 paper, it would cover 750,000 pages. However, having the genome is one thing. Understanding it, and the consequences of what it codes for, is something completely different.

The easy part is the genes. If the genome is a single long list of 3 billion letters, a gene is a short stretch of those letters that specifies how to make one or more versions of a type of molecule called a protein. If the genome is the description of the parts, the proteins are the parts themselves – solid, three-dimensional, functional. Each cell has just two copies of the complete genome sequence, one from each parent – two copies of each gene, yet each gene may lead to the production of billions of copies of the protein it describes.

The description of a protein in the genome is in code, although one that was worked out soon after the structure of DNA was determined in 1953. The genome is composed of DNA, a molecule made up of just four chemical components that we refer to as the 'letters' A, C, G and T. The letters are strung together to make a long chain, where the order of the letters is the code. Proteins are also molecules composed of a single chain of chemical components. In this case, there are 20 of them, called amino acids. How to represent 20 items with 4 DNA letters? Answer: read the DNA letters in groups of three. Groups of 2 would give only 16 combinations – not enough. Groups of three give 4 x 4 x 4 combinations, i.e. 64. Sixty-four is too many, but it turns out that several triplets code for the same amino acid.

Let's consider what is happening in every cell of your body right at this moment (which is the same in all life-forms from bacteria to human beings). A conversion from base 4 to base 64 is the sort of thing we all learned to do laboriously at school and which is now easy with pocket calculators or computers. In the cell, this translation is carried out by a molecular machine, one of the true wonders of biology, which on one side reads strings of RNA letters at the rate of three groups of three per second, and on the other assembles protein sequences from the corresponding amino acids. This is the ultimate nanotechnology, atomic-level information processing. The machine (which is called the ribosome) is itself made up of more than one hundred different protein molecules wrapping a central core formed from molecules of RNA. RNA is a sister of DNA with the same four letters, but slightly different chemistry, making it a physically more flexible molecule. DNA is more rigid, which makes it more stable and thus well suited to long-term information storage. RNA's flexibility makes it more suited to making machinery, like proteins, although it also has a role as a short-term 'message', as the ribosome does not read the DNA of a gene directly, but reads RNA copies of it. You can find the instructions for each of the parts of this machine in the genome, which means that the machine is used to manufacture its own components. This results in a chicken-and-egg paradox: one could not exist without the other. The solution to the paradox is the continuous nature of life – all cells are daughters of previous cells that divided. All cells make new ribosomes continuously, but they are also born with a starter pack of working ribosomes.

The description of the role of a genome therefore needs a slight redefinition. Rather than containing the complete instructions to build an organism, it contains only the instructions to build an organism from a functional fertilized cell. It is a big difference: a genome contains information to maintain the steady state of life, not to create it. Life is a combination of genomic instructions for maintenance, combined with a complex starting state, the living cell. Today's living cells are the end product of billions of years of evolution.

Let's get back to proteins. Although a protein molecule is a linear sequence of amino acids, just as a DNA molecule is a linear sequence of its four letters, proteins are different. They fold up into solid 3-D structures, unique to each protein sequence, and it is the shape of this structure that determines what the protein sticks to - that is, what it does in the cell. This folding process, which can be observed in a test tube, is spontaneous: the specific instructions as to how each protein is to fold up are somehow contained in the sequence of amino acids itself.

Unfortunately, we do not know how to predict the shape of a protein even when we know its sequence. This is the 'protein folding problem', which has remained unsolved after more than 40 years of research. The fact that we know anything about protein 3-D structures is because it is possible to determine them experimentally. This is a very slow process, which means that we know the sequence of many proteins for which there is no known structure. However, just as the human genome was sequenced by scaling up production, so there are now 'structural genomics' projects getting started around the world to collect this next block of data. Knowing the structure of a protein is helpful in many ways. Similarities between proteins are often more obvious from their shape than their sequence. So knowing the structure can allow us to guess the function, based on the structural similarity to another protein we know. One of the most important benefits of this is being able to study proteins and their 'substrates', to help in the design of new drugs.

There is a third type of molecule in a cell apart from DNA (the information) and proteins (the machines): these are the small molecules – the 'substrates' of proteins. We think of protein, carbohydrate and fat when we eat; each is a source of raw materials that we need in order to live. Proteins that we eat are made of the same amino acids as the proteins our DNA codes for. Our cells can't use the proteins directly – they are probably damaged as a result of being cooked or the animal or plant being dead, and would, in any case, be 'foreign' (subtly different to the ones specified by our own DNA), even if they could be transported whole into a cell. However, they can be used as a source of raw materials, amino acids, which all cells need to make new proteins. Similarly, carbohydrates are made of sugar molecules, which can be used by a cell as a source of energy, burnt by combining with oxygen to make water and carbon dioxide – which is why we also have to breathe.

All these processes – breaking down proteins into amino acids, breaking down carbohydrate into sugar and burning it for energy – take place at the molecular level and involve the protein machines. Think of a chemical plant carrying out a series of processes to convert, say, crude oil into petrol. Crude oil can be turned into lots of different things. The way to produce petrol rather than all the other possibilities is by controlling the environment: ensuring the appropriate temperature, pressure and additives at each point in the process. It's the same in the cell, but here the environment necessary for each step in each process is controlled by a different protein, through it 'binding' specifically to the molecules involved – its 'substrates'. For example, one protein might bind one molecule of sugar and one molecule of oxygen, bringing them together in exactly the right way to ensure that the reaction to release water, carbon dioxide and energy takes place, and no other. Binding between particular substrates and proteins is achieved rather in the way that keys fit into a lock –the shapes must fit exactly. This is why discovering the 3-D structures of proteins is so important, since observing the shape of the pockets that bind the substrates helps in the selection and design of other small molecules that might bind there, such as pharmaceutical drugs.

Most drugs help by either blocking or stimulating one of the natural processes in your body, or by fatally blocking one of the processes of a foreign organism that is attacking you, such as a bacteria or virus. When drug companies talk of 'targets', this is what they mean: the protein machines that drugs are aimed at. Our knowledge of small molecules also leads to something else: the organization of proteins according to their roles in different cell processes. Just as in any manufacturing process, cellular processes, such as the transformation of food into energy and components to make more proteins, have many steps and involve many proteins. The first protein aids the conversion of molecule A into B, the second protein aids the conversion of molecule B into C, and so on. These processes are referred to as pathways and many have been worked out over the years, before we got the genome sequences, by working with proteins directly.

So, if we consider what we know about DNA, proteins, small molecules and their organization, we may feel that we already have a pretty good picture of how a cell works. We have the complete code written in the genome as DNA; we have a partial list of proteins in the cell and a partial list of the processes that they are involved in. We can expect continuing experimental work to fill in the gaps in our knowledge of the proteins and their 3-D structures and of the processes they are involved in. However, there is something crucial missing.

Think about the cistern of your toilet. The toilet flushes and the cistern refills until it is full, at which point it stops. The level of water controls the valve, allowing it to remain open until there is enough water, at which point it is closed. This mechanism is a 'feedback loop': the water acts to stop there being too much water. Now think of a gene specifying how to make a protein. How can it make sure that not too much protein is made? It turns out that the region of DNA that codes for a protein – a gene – has nearby 'on-off' switches. These are extra letters of DNA that particular proteins bind to (similar to the way they bind small molecule substrates), and which control whether the gene is turned on or off. In many cases, the control is provided by the protein that is being made. The protein binds itself to the 'control' letters of the gene and stops more protein being made. This is a feedback loop.

How does this actually work? You might worry that since only a single molecule of protein needs to binds to the control region to turn the gene off, only one molecule of anything would ever get made. Fortunately, biology doesn't work like that because of physics. Everything around us might seem solid, but at the atomic level all the atoms and molecules are moving. It is the same in a cell, which is like an ocean of molecules. Think for a moment of genes as if they were islands in that ocean. Make a molecule of protein and it's a bit like dropping a bottle into the ocean. That

A cell is like an ocean of molecules. Think for a moment of genes as if they were islands in that ocean. Make a molecule of protein and it's a bit like dropping a bottle into the ocean

protein might contain in its shape a 'message' to turn the gene that made it off, but in order for that message to work, it must wash up on the shore of the gene island where it was made, that is bind to the gene's control sequence. Suppose it floats off into the ocean and is never seen again? This happens – protein molecules get made until there are so many that some will randomly wash up on the shore of the gene that made them and turn production off. And if they can wash up on a shore, they can wash off too (i.e. unbind from the control sequence). In this way the fraction of time that a control region has a protein bound to it reflects the number of molecules in the cell as a whole. It is this transient binding to the DNA of the gene by its own protein product that enables it to 'regulate' its own production.

Of course, the cell is full of lots of different types of protein, all bumping into the control regions of different genes. However, this is not a problem. The binding between a protein and its gene is specific, just as proteins bind specific substrates when carrying out their production function. For binding substrates, the lock and key analogy is appropriate, but for regulation where different large molecules interact, it is perhaps easier to think of proteins as being sticky, but in a way that each only sticks to certain things. Consider the two components of Velcro – neither sticks to anything else, but they stick to each other strongly. Examine the components closely and you see a fine structure of complementary hairs and hooks. This is rather like the surfaces of proteins that must stick together.

We know a reasonable amount about protein structures and their substrates partly because of scale – you don't live off the energy of one molecule of sugar but of trillions. There are enough of these types of molecules in a cell to allow a pure sample to be extracted so that the 3-D structure can be studied. The stickiness involved in regulation is a different matter. There are only two copies of a gene per cell, making it difficult to isolate enough of what binds to a gene sequence in order to study what it is, especially with the binding being transient. This is the missing part of the puzzle. We can guess that many proteins may regulate their own production, but this is just the tip of the iceberg of control mechanisms of the cell, let alone an organization of cells that makes a body. The study of all the proteins encoded by a genome, their function, structure and interactions, has been loosely named 'proteomics', and is being advertised as the next big project. However, whereas the job of sequencing DNA of a genome is clearly finite, studying proteins and all their interactions is much less well defined and much harder.

It might be difficult to observe the individual interactions between molecules, but we are at least beginning to be able to observe the ways in which genes and proteins work together. The gene chip is a tiny glass plate with a fragment of the complementary sequence of every gene in a cell attached to a different position on a grid. In any cell, only some genes are active and for these genes, the cell will contain RNA copies of the DNA sequence (the molecules that act as 'messengers' between the genome and the ribosome). First, take the contents of a cell and chemically attach a fluorescent dye molecule to each molecule of RNA. Then wash the RNA over a gene chip and each molecule will stick to its complementary sequence. Since the complementary sequences are attached on a grid and the molecules that stick to them are fluorescent, you can identify all the active genes by observing which grid points light up. The amount of RNA that is bound can be measured by the light intensity, and hey presto, you have a snapshot of exactly which genes were turned off and an estimate of how much protein was being made from genes that were turned on – information that was previously unavailable.

There is more. Compare the results for a normal cell with those for a cancer cell and you see instantly how the pattern of protein production has changed – which genes are more active, which are less active. This immediately gives us a tool that can aid in the diagnosis of disease, since it is likely that each specific pattern of change can be correlated with a particular disease. Gene chip diagnostics will be one of the first major benefits of the genome project.

Gene chips can show us something of what has happened in a cell, but not why or how. If we are going to understand processes like cancer, the latter is important. Part of what makes this difficult is that the cell is an integrated system where almost everything affects everything else. There may be so many interactions that it becomes hard to work out what is going to happen by observing the external behaviour. We can even begin to see this behaviour in man-made machines. Machines are made of self-contained components, which may individually have a well-defined behaviour, but which may attempt to compensate for changes in other components.

The result is that a failure may be hidden for a while, or perhaps even completely compensated for. Although machines cannot repair themselves (yet), this compensation behaviour is not unlike the way our bodies and our cells can cope, most of the time, with infection, injury and ageing.

Consider something like the temperature gauge on your car. You notice the temperature of your car engine starting to rise. Is there a water leak? Did you forget to fill it with oil? Or is it just that you are driving uphill in a traffic jam on a hot day? Now compare this to your body. Why did you suddenly lose weight last month? Was it stress at work, your change in diet or something more serious?

Working out what is really going on when some sort of sickness is suspected is not obvious even in machines. At least with modern cars, a garage may be able to access a log kept by the car's computer of everything that has gone on, as well as a snapshot of the state of many components. The gene chip is likely to become the first equivalent tool in component-level diagnosis of a human body, making it possible to see some of what has changed at a molecular level. It will not only be the first practical product to come out of the human genome project, but also the first to significantly change the direction of medicine.

Machines are the same, but people differ. These are not the differences of experience that cause us to have different minds, but the differences in our genome sequence that affect the way our human body machine works. Our genomes are remarkably similar across the entire planet. Yet there are several million of the 3 billion positions where alternative letters are commonly found. We each have a unique combination of these alternatives – a personal genome – and some of these alternatives will affect the way a gene or its protein product behaves. If one such alternative makes a gene appear to work less well, it does not mean the effect is necessarily serious or even that it does not have some hidden advantages.

Again, think of a cistern. Some valves never close properly, with the result that the water never stops completely. So cisterns are designed with an overflow. It is not ideal to have water dripping from an overflow, but it is better than a flooded bathroom. On the other hand, a drip would keep the water in the supply pipe flowing slowly and stop the pipe freezing, which could be an advantage. Similarly, a difference in a single letter of your DNA could result in a gene never being completely turned off, which results in slightly too much of that protein being made. With 30,000 genes, the chance of a difference like this affecting one of them is quite high, so cells have safety systems too – the equivalent of overflows – to make sure that quite large differences in protein production can be compensated for, perhaps not perfectly, but at least well enough not to be fatal. It might not be perfectly efficient, but in a complex system it becomes necessary to assume that things will go wrong and will have to be compensated for. Evolution selects for advantage, so alternatives that exist widely in a population are either advantageous in some situations now or used to be advantageous in the past, and have not yet been bred out of the population.

Some alternative sequences of a gene, however, may turn out to have important consequences, even if only in rare man-made situations. Your version of a gene appears to behave normally, but if you take a drug, it works in a different way. It could be that for you, the drug gives side effects or it does not work at all. Drugs are known to work for only some of the people they are used to treat and to generate side effects in others. If we know the genome and its potential differences, it will be possible to relate this to the behaviour of each drug. This will soon lead to personalized medicine. The doctor will check the appropriate sequence of your genome and prescribe the drug that works for you. Eventually there will be new drugs for people for whom existing drugs do not work, because of their gene variations.

This may seem like fiddling at the edges of the problem of health, but to go beyond this is much harder. Hardly a day seems to go by without a new report of a gene that has been identified as affecting some disease. Unfortunately, it is rarely a single change at one point in the genome that causes a disease. When two people have different risks of heart disease, this is likely to be due to subtle combinations of differences in their genome at many different places, involving many different genes. If we want to understand this, we are probably going to have to really understand the body as a machine. It is time to think again about complexity.

Until recently, aeroplane designs were tested in wind tunnels. This was because their behaviour could not be predicted. If a lever in the cockpit is directly connected to the tail, the plane shakes violently when you move that lever. So levers cannot be connected directly – the mechanism needs to be more complex, with 'feedback' from the part being moved, rather like the way a protein feeds back to control the gene that makes it. As soon as this complexity is introduced, the system becomes non-linear. We are unable to model non-linear systems easily on computers. Once a system is non-linear, it is hard to anticipate all the ways in which it might behave. Does the mechanism always work or does it at some point behave unpredictably?

The non-linear systems we currently design are very simple. And it is clear from computers that we are not very good at building robust complex systems, since computers crash, unexpectedly – there are too many interactions between the parts of the software system to anticipate and test for all the things that could happen. In cells, there are tens of thousands of different types of molecules randomly colliding with each other, and yet there is no evidence that cells die as a result of internal breakdowns. The cell is just the lowest level in the system of the human machine. The body is highly complex and doesn't crash. When you think that our bodies are constructed through a series of programmed steps, beginning with a single cell that divides no more than 50 times to result in an organized structure of 100 million million cells, in an environment that is different in every case, the amazing thing is that the process ever works. Complexity systems theorists now regard biology as the only proof in existence that large complex systems can be constructed. How did biology evolve this solution and when will we understand it?

It is clear that biology is a curious mixture of information and physics. We can easily process the raw information contained in the genome sequence on a computer, whereas the cell has to make amazingly complex machines to read DNA and interpret it as a sequence of amino acids. On the other hand, 40 years of research have not solved the protein folding problem, despite the use of the fastest computers. In cells, there is no machinery for protein folding – biology relies on the laws of physics to lead a protein chain to spontaneously find a single, stable, 3-D structure. We may find a short cut by working out the structures of most proteins experimentally, rather than solving the problem mathematically, but predicting the stickiness between specific molecules requires exactly the same understanding as the folding problem. Working out all the interactions in a cell between tens of thousands of objects is going to be much harder to address experimentally. And if we end up with the full list of interactions, like the list of interacting parts in a car, will we be able to model how they work together? Just two components can interact in ways to produce unpredictable effects. A single cell, made up of tens of thousands of interacting components, is infinitely more complex than any system humans have designed. Even the mechanics of the control systems of one of the simplest viruses, sequenced decades ago, are still not fully understood.

Looked at like this, building accurate models of cells and organisms on computers might seem hopeless. However, the very fact that biological systems work with such great reliability implies that they must accommodate variations in almost everything. This includes variations between different species, since most biological components, whether at the level of individual cell-types or large parts of a body like a heart, are built in the same way in any organism that has them.

This suggests that control mechanisms that allow cells and bodies to be constructed and maintained must be made up of units of robust logic, each of which will produce the same behaviour over a range of different environments.

If we know the genome and its potential differences, it will be possible to relate this to the behaviour of each drug. This will soon lead to personalized medicine

It may even be appropriate to make the analogy with object-orientated programming in computing, which for the same reason attempts to divide computer programs into units of well-defined, self-contained code – to enhance the robustness and maintainability of very large systems. Even if we are unable to measure all the parameters of each system, such as exactly how strongly any two proteins interact, we may still be able to deduce the logical behaviour of each system, simply because each must have one. It may then be possible to build up a useful description of the human machine from these logical units, which is predictive insofar as we can use it to anticipate the consequence of drugs or even the individual differences in our genome sequence. Increasingly, however, it seems that the evidence required to make these deductions is going to come from painstaking biochemistry and molecular biology. Sequencing the genome and other large-scale surveys of cellular behaviour, such as by gene chips, is giving us huge global blocks of data, but will provide only part of the story.

The very presence of these large data sets is having an enormous impact on biology. It has begun to put numbers to all the genes and all the proteins. Furthermore, all data has something to say about biology: it can all be linked to genome sequences. And all genome sequences, which are related to each other through evolution, can be linked together too. This means that potentially all information in biology could help us understand ourselves. And the amount of data is huge – biological databases have grown faster than both the growth in computer speed or disk storage and now require large computer centres to handle them, even challenging the domination of physics as a user of high-performance computing. The term 'bioinformatics' has been coined, but it is not just a term for specialists. It is almost impossible to be a molecular biologist today and not use these bioinformatics resources.

Similarly, collecting and organizing individual research results from thousands of laboratories worldwide into these databases, alongside blocks of data produced on a large scale, has become an industry in its own right. It is no longer enough for results to be published only in scientific journals. Much of the understanding only comes when many independent results are considered together, as part of a system, and for that they need to be organized in databases. Biology is not new, but the integrating power of genomes means that understanding the human machine has now become a project that involves all biologists, all over the world.

Fountains of life

Tim Radford

In 2001 Britain became the first state to formally permit the use of embryo stem cells for research

Stem cells are the link between a fertilized egg the size of a full stop and the fully-formed tiny human being that appears nine months later. An embryo begins as one cell and turns into a baby composed of trillions of specialized cells of more than 200 different kinds. This cascade of variation and renewal begins with a tiny ball of embryo stem cells.

Britain entered the history books in January 2001 by becoming the first state to formally permit the use of embryo stem cells – and by implication, embryos – for research into human diseases. It required a change of the rules governing the use of embryos up to 14 days old, left over from infertility treatment. At the end of two years of consultation, debate, committees and reports, the change finally went through the House of Lords on 22 January. Some biotech campaigners, researchers and medical charities were delighted; other groups were appalled. But the hope is that embryo stem cells could be cultivated and used as the basis for cures for neurodegenerative conditions for which no treatments exist, such as Parkinson's disease. Other future candidates for stem-cell based treatments could be hepatitis, leukaemia, diabetes and rheumatoid arthritis.

The interest in stem cells grew out of the famous experiment with Dolly the cloned sheep in Scotland in 1997, and the isolation of human embryonic stem cells by a US group in 1998. The first showed that the biological clock could be turned backwards: that an 'adult' cell could become the genetic begetter of a new baby. The second showed that embryo stem cells could be isolated and kept alive. The two discoveries immediately seemed to open a new door for medicine. Stem cells could provide new tissue for failed livers or heart muscle, or damaged spinal cords. The problem is that stem cells are difficult to identify – brain stem cells look just like ordinary brain cells, skin stem cells look like any

A flask of human embryo stem cells in Jim McWhir's laboratory at the Roslin Institute where Dolly the cloned sheep was created.

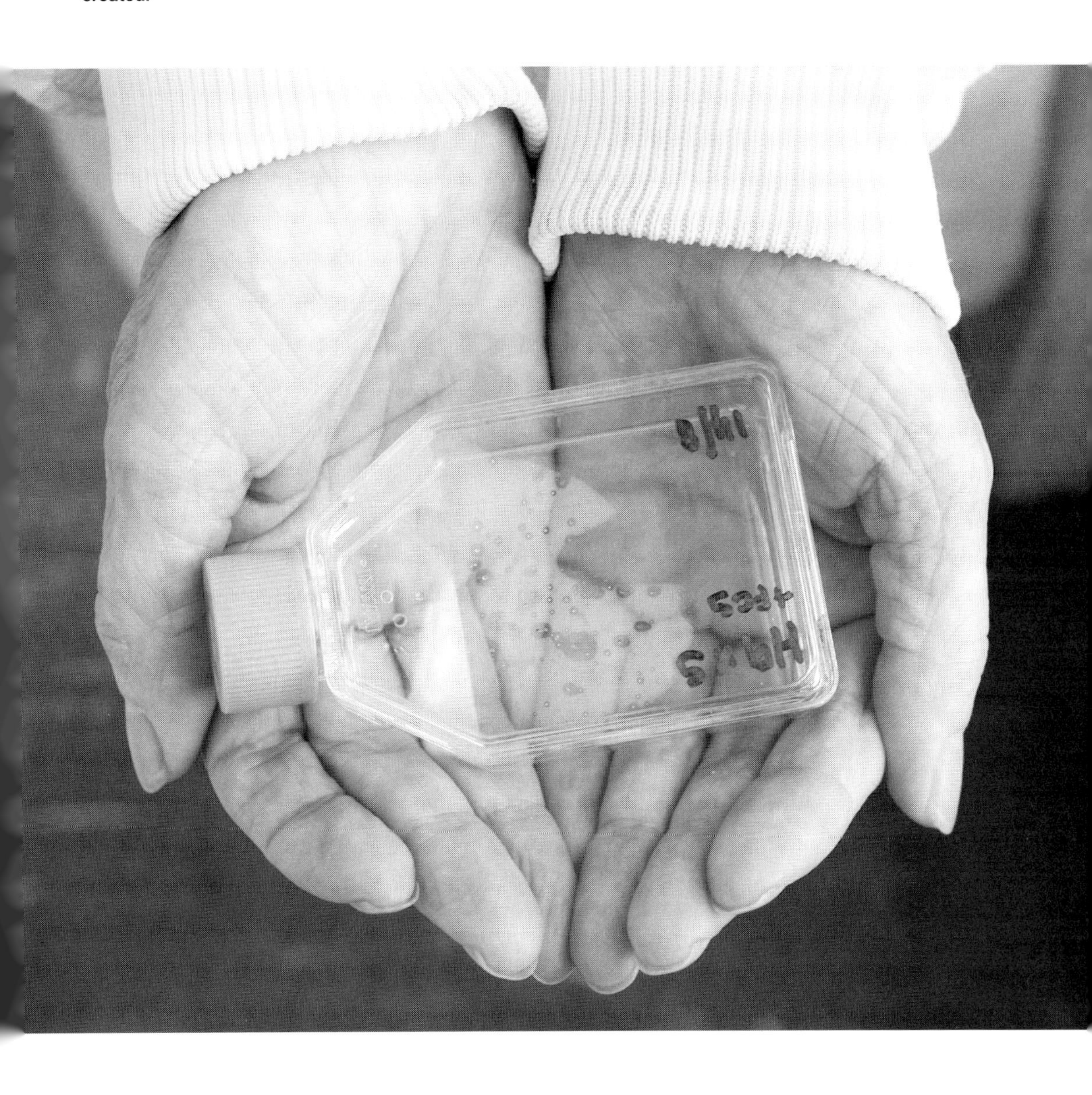

other skin – and it could be slow work growing enough 'personalized' stem cells to replace tissue in a failing patient. The Royal Society proposed that researchers should start collecting banks of stem cells that might one day be matched with tissue type and used to repair or even replace diseased organs.

Meanwhile, the supply of embryos donated for research and for future treatments will almost certainly be limited. Initially, at least, some researchers dreamed of cultivating 'spare parts' from embryo stem-cell cultures, a technique referred to as therapeutic cloning. There was uproar in November when Advanced Cell Technology of Massachusetts announced that it had successfully cloned human embryos as part of its research into stem-cell treatments. But the embryos failed when they were too tiny even to be a source of stem cells. Trials of stem-cell based treatments on human patients are not likely for several years. Many people – some within the biological sciences – felt uneasy about the use of embryos for anything except fertility research, and argued that adult rather than embryo stem cells might be turned into medical tools.

Right on cue, such things began to seem possible in 2001. In March, two teams in New York separately reported that they took stem cells from adult bone marrow and turned them into heart tissue. One team injected stem cells into mice with damaged heart tissue: nine days later, new cells were growing in 68 per cent of the injured ventricle. The other group injected human stem cells into the tails of rats recovering from a heart attack. When they looked, they found that new human-type capillaries had formed in damaged parts of the heart. A group from Florida reported that if umbilical cord blood, a source of stem cells, was injected into the veins of stroke victims, it could perhaps restore damaged brains. Once again, it worked in rats.

In April, a team from the University of California, Los Angeles, reported that they had found a way to take fat cells and turn them into bone, muscle and cartilage for transplant surgery. 'Stem cells are like kids who, when they grow up, can enter a variety of professions,' said Marc Hedrick of the UCLA School of Medicine. 'Fat is perhaps the ideal source. There's plenty of it.'

A group from Florida reported that if umbilical cord blood was injected into the veins of stroke victims, it could perhaps restore damaged brains. It worked in rats

The genome harvest

Tim Radford

The human genome was just one among many published by the genetic sequencers in 2001. The European agribusiness giant Syngenta and Myriad Genetics of Utah – which holds the patents on two breast cancer genes – claimed to have completed the entire genetic code of rice, the crop that feeds half the planet. They completed the task far ahead of a publicly funded consortium, triggering alarm from Action Aid, the hunger charity, which pointed out that the biotech companies already held 229 patents on rice, the diet of the poorest 3 billion people in the world.

But British teams – and their international partners – claimed and published in 2001 the entire DNA 'texts' of a number of other useful organisms. One was the typhoid microbe *Salmonella typhi*. Another was the sinister bacterium *Yersinia pestis*, better known as bubonic plague or the Black Death. This has a special place in history: it is believed to have wiped out one third of Europe's population during the 14th century, and the World Health Organization still records up to 3,000 cases a year. The strain sequenced by British scientists was taken from a vet in Colorado, who died in 1932 after rescuing a cat from underneath a house. The cat sneezed on him and he caught the pneumonic and most lethal form of the plague. 'Many people do not realize the plague is still with us,' said Julian Parkhill of the Wellcome Trust Sanger Institute in Cambridge. 'There are even some drug-resistant strains in Africa. It is down but not out, which is why this report is so important.'

Later in the year, an 11-nation consortium led by scientists in Rome, Italy, announced they were going to sequence the genetic code of the Musaceae family of plantains and bananas, on which at least 500 million people depend. 'Bananas are a staple food that many African families eat at every meal,' said Emile Frison of Inibap, the international network for the improvement of the banana and the plantain, based in Montpellier, France. 'This is our chance to develop a crop that won't fail for them, and may help lift them out of hunger and poverty.'

A space of one's own

Martin Rees

Why is the universe a fit place for life? How will the universe end? And is this the only universe?

Early in the 20th century, astronomers realized that our galaxy was just one of billions. Later, new concepts and techniques vastly extended the range and richness of observed cosmic phenomena: quasars, black holes, neutron stars and the 'Big Bang' entered the general vocabulary, if not the common understanding. Cosmologists can now set our entire solar system in a grand evolving scenario stretching back to the first seconds after the Big Bang – an era when everything was squeezed into an amorphous gas hotter than the centres of stars. We have taken the measure of the universe, just as in earlier centuries navigators mapped the Earth's oceans and continents. We are witnessing a crescendo of discoveries that promises to continue in the new millennium.

In 1965, Arno Penzias and Robert Wilson famously detected the 'afterglow' of the hot, dense beginning of our universe. Evidence for the Big Bang has accumulated and become far more precise in the succeeding 35 years. Some debates have been settled; some issues are no longer controversial. New cosmological questions that couldn't even have been posed in earlier decades are now being debated – in particular, cosmologists are trying to probe back still further, to understand why the Big Bang occurred in the way it did.

The simple recipe that describes our one-second-old universe – the proportions of different kinds of particles and radiation, the cosmic expansion rate, and so forth – must be the outcome of what happened still earlier: within the first tiny fraction of a second. The conditions that prevailed then are more extreme and less familiar. For the first few microseconds, everywhere would be denser than an atomic nucleus. Experimenters at the CERN laboratory in Geneva and at Brookhaven National Laboratory in the US have replicated these conditions on a tiny scale by crashing together nuclei of lead and gold accelerated to almost the speed of light to probe their constituent quarks. Further back still, the energies and densities are so extreme that experimenters have even less of a foothold. In the first trillionth of a second, each particle would have carried more energy than the most powerful accelerators can reach.

But we probably have to extrapolate even further back in order to answer one of the most fundamental questions: why is the universe expanding the way it is? It is seriously misleading to think of the Big Bang as being triggered by an explosion. Bombs on Earth, or supernovae in the cosmos, explode because a sudden boost in internal pressure flings the ejecta into a low-pressure environment. But in the early universe, the pressure was the same everywhere: there was no empty region outside. We need some other explanation for what banged and why it banged. According to the best current guess, the universe started off as an infinitesimal speck. It underwent a period of 'inflationary' expansion, driven by energy latent in space. The expansion was exponential; the scale doubled, then doubled, and then doubled again. Within about a trillionth of a trillionth of a trillionth of a second, it is claimed, an embryo universe could have inflated large enough to encompass everything we now see. And then, the fierce repulsion switched off; some of the energy converted into heat, initiating the more familiar expansion process that has led to our present habitat.

Inflation stretches a microscopic patch until it becomes large enough to evolve into our observable universe. Our universe then ends up being 'stretched flat', rather as any part of a wrinkled surface becomes smooth if it is stretched enough. The process would be likely to overshoot, inflating by far more than is needed to account for the 10-billion-light-year dimensions of our observable universe: the distance to the edge could be a number with millions of zeros. In this expanse of space, far beyond the horizon of our observations, the combinatorial possibilities are so immense that close replicas of our Earth and biosphere would surely exist, however improbable life may be. Indeed, in a sufficiently colossal cosmos there would, somewhere, be exact replicas not just of our Earth, but also of the entire domain (containing billions of galaxies, each with billions of stars) that lies within range of our telescopes.

Even this stupendous expanse of space may not be everything there is. Some theories suggest that our Big Bang wasn't the only one. Patches where inflation doesn't end might always grow fast enough to provide the seeds for other big bangs. This line of speculation dramatically enlarges our concept of reality – from a universe to a multiverse. This scenario might make some features of our own cosmic habitat less surprising.

The assertion that everything inflated from something microscopic looks like 'something for nothing', but it isn't really. Our present vast universe may, in a sense, have zero net energy. Every atom has an energy because of its mass – Einstein's mc^2. But it has a negative energy due to gravity – we, for instance, are in a state of lower energy on the Earth's surface than if we were up in space. And if we added up the negative potential energy we possess due to the gravitational field of everything else, it could cancel out our rest mass energy. Thus it doesn't, as it were, cost anything to expand the mass and energy in our universe. This concept of 'inflation' isn't just metaphysics. Its one generic prediction – that the universe would be stretched 'flat' – seems to be gratifyingly borne out by observation. Measurements of structures in the heavens are measures of triangles, and these triangles indeed have angles adding up to 180 degrees, even if their sides stretch to the limits of the observable universe.

Is the expansion slowing down or speeding up?

So much for the past. What can cosmologists infer about the distant future? The sun is less than halfway through its life. It will continue to shine for five billion years – several times longer than it has taken for Earth's biosphere (including us) to evolve from the first multicellular organisms. The entire galaxy, extending for 100,000 light years, could be 'greened' in less time than it took for us to evolve from the first primates. There is plenty of time for life to spread. Even if life is now unique to Earth it could eventually take over the cosmos.

We are far from the culmination of evolution – the emergence of structure, intelligence and complexity is still near its cosmic beginnings. The future may be determined not by natural selection, but by human decisions, or by artefacts, created by us, that develop via their own directed intelligence. Diffuse living structures, freely floating in interstellar space, would think in slow motion, but may nonetheless come into their own in the distant future. Our universe has the potential to harbour a teeming complexity of life far beyond what we can even conceive.

Life's long-range future will long remain highly speculative, but we can already forecast the fate of stars and galaxies. When the sun dies, the galaxies will be more widely dispersed, and will be intrinsically somewhat fainter because their stellar population will have aged. But what might happen still further ahead? Will the universe go on expanding for ever? Or will our firmament eventually collapse again in a Big Crunch? Space is already being punctured by the formation of black holes, but is this just the precursor of a crunch that will engulf us all?

The answer depends on how much the cosmic expansion is being decelerated. Everything exerts a gravitational pull on everything else; and it is

There is plenty of time for life to spread. Even if life is now unique to Earth it could eventually take over the cosmos

easy to calculate that it would take only the mass of five hydrogen atoms per cubic metre to bring the expansion to a halt – unless some other force intervenes. This doesn't sound like much. But if all the galaxies were dismantled, and their constituent stars spread uniformly through space along with all the gas, they'd make an even emptier vacuum – one atom in a volume of ten cubic metres. There seems to be a similar amount of material in diffuse intergalactic gas, but even when that is added, the resulting density amounts to only 0.2 hydrogen atoms per cubic metre. That's equivalent to a few grains of sand in the volume of the Earth – or just one small asteroid, a few hundred metres across, in a box big enough to contain our entire solar system. And it is 25 times less than the critical density of five atoms per cubic metre. This may seem to imply perpetual expansion, by a wide margin. But the actual situation is less straightforward, because of the mysterious dark matter. Galaxies, and even entire clusters of galaxies, would fly apart unless they were held together by the gravitational pull of five to ten times more 'dark' material than we actually see.

Dark matter, though it outweighs ordinary atoms, still contributes no more than about 30 per cent of the density needed to halt the expansion – in cosmological jargon, the value of omega, the ratio of the actual density to the so-called 'critical density', is only about 0.3. This finding suggests that the universe is not slowing down enough to ever come to a halt.

There is another approach to this long-range forecasting, which at first sight seems more straightforward: namely, to look directly for a difference between the expansion rate a few billion years ago and the present rate, and then extrapolate the trend forward. This comparison is possible in principle because the red shifts – light shifts its frequency according to the speed of its source rather as the noise of a passing train seemingly changes pitch – of distant objects tell us how they were moving when their light left them, not how they're moving now. The best standard candle yet recognized is a type of supernova that is triggered by a nuclear explosion. These so-called Type 1A supernovae are, in effect, thermonuclear bombs, where a particular type of star exploded, with a standardized yield. The first results using this technique – announced in 1998 by two international teams of researchers – caused a stir. There certainly did not seem to be as much deceleration as would be expected in the simplest kind of universe where omega was exactly one. That in itself was not surprising. What was a surprise was that the expansion seemed actually to be speeding up.

The magazine *Science* rated this finding as the number one scientific discovery of 1998, in any field of research. Acceleration implies an extra cosmic force – some kind of cosmic repulsion that overwhelms gravity. This idea goes back to Einstein in 1917. At that time, astronomers only really knew about our own galaxy – not until the 1920s did a consensus develop that Andromeda and similar 'spiral nebulae' were actually separate galaxies, each comparable to our own. It was natural for Einstein to presume that the universe was static, neither expanding nor contracting. He found that a universe could not persist in a static state unless an extra force counteracted gravity. He incorporated an extra number into his equations, which he called the cosmological constant, denoted by the Greek letter lambda. This introduced a repulsive force – a kind of 'antigravity' – that counterbalanced the ordinary gravity and allowed a static universe that was finite but unbounded. In Einstein's static universe, any light beam you transmitted would return and hit the back of your head.

Einstein, in his later life, rated lambda as his 'biggest blunder', because if he had not introduced it to permit a static universe he might have predicted the expansion before Edwin Hubble's discovery in 1929. Einstein's motive for inventing lambda has been obsolete for 70 years; but that doesn't discredit the concept itself. On the contrary, lambda now seems less ad hoc than he thought it was. Lambda can be envisioned as energy somehow contained even in empty space ('vacuum energy', as physicists put it). According to our present concepts, empty space is anything but simple. All kinds of particles are latent in it. On an even tinier scale, it may be a seething tangle of strings. From our modern perspective the puzzle is not why there should be a lambda: it's why it is not much, much higher.

If there is energy in empty space (equivalent, as Einstein taught us, to mass, through his famous equation $E=mc^2$), why does it have the opposite effect on the cosmic expansion from the atoms, the radiation and the dark matter, all of which tend to slow down the expansion? The answer depends on a feature of Einstein's theory that is far from intuitive: gravity, according to the equations of general relativity, depends not just on energy (and mass), but on pressure as well. And a generic feature of the vacuum is that if its energy is positive, then its pressure is negative (in other words, it has a 'tension', like stretched elastic). The net effect of vacuum energy, then, is to accelerate the cosmic expansion. It has a huge negative pressure and so, according to Einstein's equations, it pushes rather than pulls.

Just within the last two or three years, cosmologists have developed a consensus about the basic ingredients that make up our universe. Ordinary atoms, in stars, nebulae, and diffuse intergalactic gas, provide just 4 per cent of the mass; dark matter provides 20–30 per cent; the rest (i.e. 66–76 per cent) is dark energy latent in empty space. The expansion accelerates because dark energy (with negative pressure) is the dominant constituent. In still earlier centuries, the classical view was that everything in the 'sublunary sphere' consisted of the four 'elements' – earth, air, fire and water – but the heavens were constituted from some quite different 'fifth essence'. This concept was laid to rest in the 19th century, when studies of stellar spectra showed that stars were made of the same stuff as the Earth. Modern cosmology revives a similar antithesis. It looks as though the mysterious 'antigravity substance' – vacuum energy – provides the dominant mass-energy in our universe, even though it plays no role in stars or galaxies, and that ordinary atoms are just an 'afterthought' or minor pollutant in a cosmos whose overall evolution is governed by quite different substances.

We can never be quite sure of the long-term cosmic future. But the odds favour perpetual expansion: the gravitational pull of all the atoms and dark matter in the universe is insufficient to bring cosmic expansion to a halt. Events far too rare to be discernible today could gradually come into their own – stellar collisions, for instance. Stars are so thinly spread through space that collisions between them are immensely infrequent (fortunately for our Sun), but their number would mount up. The terminal phases of galaxies would be sporadically lit up by intense flares, each signalling an impact between two dead stars. Eventually, even black holes will decay. The evaporation of black holes, being a quantum process, is far less important for big holes: the time it would take for a hole to erode away depends on the cube of its mass. The lifetime of a hole with the mass of a star is 10^{66} years. Even black holes as heavy as a billion suns – such as those that lurk in the centres of galaxies – would erode away in less than 10^{100} years.

The asymptote for life

Cosmologists have produced an immense speculative literature on the ultra-early universe: thousands of papers have, for instance, been written just on the concept of 'inflation'. In contrast, cosmic futurology has largely been left to science fiction writers. But 20 years ago, one distinguished scientist, Freeman Dyson, helped to make the subject scientifically respectable: he published a fascinating and detailed article in *Reviews of Modern Physics* called 'Time without End: Physics and Biology in an Open Universe'. The evidence for an ever-expanding universe was then less clear than it is now. But Dyson already had his prejudices: he wouldn't countenance the Big Crunch option because it 'gave him a feeling of claustrophobia'. He discussed the prognosis for intelligent life. Even after stars have died, he asked, can life survive for ever without intellectual burn-out? Energy reserves are finite, and at first sight this might seem to be a basic restriction. But he showed that this constraint was actually

Will the universe go on expanding for ever? Or will our firmament eventually collapse again in a Big Crunch? Space is already being punctured by the formation of black holes, but is this just the precursor of a crunch that will engulf us all?

not fatal. As the universe expands and cools, lower-energy quanta of energy (or, equivalently, radiation at longer and longer wavelengths) can be used to store or transmit information. Just as an infinite series can have a finite sum (for instance $1 + 1/2 + 1/4 + ... = 2$) so there is no limit to the amount of information processing that could be achieved with a finite expenditure of energy. Any conceivable form of life would have to keep ever cooler, think slowly, and hibernate for ever-longer periods. But there would be time to think every thought. As Woody Allen once said: 'Eternity is very long, especially toward the end.'

Dyson imagined the endgame being spun out for a number of years so large that to write it down you'd need as many zeros as there are atoms in all the galaxies we can see. At the end of that time, any stars would have tunnelled into black holes, which would then evaporate, in a time almost instantaneous in comparison. In the 20 years since Dyson's article appeared, our perspective has changed in two ways, and both make the outlook more dismal. First, most physicists now suspect that atoms don't live for ever. In consequence, white dwarfs and neutron stars will erode away, maybe in 10^{36} years. The heat generated by particle decay will make each star glow, but as dimly as a domestic heater. By then our Local Group of galaxies (those nearest our own) would be just a swarm of dark matter and a few electrons and positrons. Thoughts and memories would only survive beyond the first 10^{36} years if downloaded into complicated circuits and magnetic fields in clouds of electrons and positrons – maybe something that would resemble the threatening alien intelligence in *The Black Cloud*, the first and most imaginative of Fred Hoyle's science fiction novels, written in the 1950s.

Dyson was optimistic about the potentiality of

an open universe because there seemed to be no limit to the scale of artefacts that could eventually be constructed. He envisioned the observable universe getting ever vaster, but gradually slowing down in its expansion. But now we know that gravity is overwhelmed by a cosmic repulsion, causing galaxies to move apart at an accelerating rate. This realization makes the long-term future more constricted. Galaxies will fade from view even faster: they get more and more red-shifted – their clocks, as viewed by us, seem to run slower and slower, and freeze at a definite instant, so that even though they never finally disappear we would see only a finite stretch of their future. The situation is analogous to what would happen if a cosmologist fell into a black hole: from a vantage point safely outside the hole, we would see them freeze at a particular time, even though they would experience, beyond the horizon, a future that is unobservable to us. Our own galaxy, Andromeda, and the few dozen small satellite galaxies that are in the gravitational grip of one or other of them, will merge together into a single amorphous system of ageing stars and dark matter. In an accelerating universe, everything else disappears beyond our horizon; if the acceleration is due to a fixed 'lambda', this horizon never gets much further away than it is today. So there is a firm limit – though of course a colossally large one – to how complex any network or artefact can ever become.

A biophilic universe?

An iconic image from the 1960s was the first photograph of the whole Earth from space: the fragile beauty of our home planet's land, oceans and clouds contrasted starkly with the sterile moonscape on which the astronauts left their footprints. We have learned in the last five years that stars are not mere 'points of light' – many are other Suns, orbited by retinues of planets. So far, the only detectable extra-solar planets are giant ones, the size of Jupiter or Saturn, but within 20 years we will be able to hang on our walls another poster that will have even more impact – a telescope image of another Earth, orbiting some distant star. But will this alien planet have a biosphere? Is it inevitable, or even highly probable, that 'simple' organisms would emerge on a planet whose size and orbit resembled the Earth? Even if they did, what is the chance that they would evolve into something that could be called intelligent?

If we ever established contact with intelligent aliens, how could we bridge the 'culture gap'? One common culture would be physics and cosmology. We would all be made of atoms, and we would all trace our origins back to the same 'genesis event' – the so-called Big Bang, which happened about 13 billion years ago. We would all share the potentialities of a (perhaps infinite) future. But we (and the aliens, if they exist) would realize that a universe that could harbour life, or indeed any complexity, must be rather special. This realization brings within the framework of science some fundamental questions that were formerly in the realm of speculation.

Any universe hospitable to life – what we might call a biophilic universe – has to be 'adjusted' in a particular way. The prerequisites for any life – long-lived stable stars, stable atoms such as carbon, oxygen and silicon, able to combine into complex molecules, etc – are sensitive to the physical laws and to the size, expansion rate and contents of the universe. If the recipe imprinted at the time of the Big Bang had been even slightly different, we could not exist. Many recipes would lead to stillborn universes with no atoms, no chemistry, and no planets; or to universes too short-lived or too empty to allow anything to evolve beyond sterile uniformity. This distinctive and special-seeming recipe seems to me a fundamental mystery that should not be brushed aside merely as a brute fact.

For example, we could not exist in a world where gravity was much stronger. In an imaginary 'strong-gravity' world, stars (gravitationally bound fusion reactors) would be small; gravity would crush anything larger than an insect. But what would preclude a complex ecosystem even more would be the limited time. The mini-Sun would burn faster, and would have exhausted its energy before even the first steps in organic evolution had got under way. A large, long-lived and stable universe depends quite essentially on the gravitational force being exceedingly weak.

The nuclear fusion that powers stars depends on a delicate balance between two forces: the electrical repulsion between protons, and the strong nuclear force between protons and neutrons. If the nuclear forces were slightly stronger than they actually are relative to electric forces, two protons could stick together so readily that ordinary hydrogen would not exist, and stars would evolve quite differently. Some of the details are still more sensitive. For instance, carbon – crucial for all life – would not be so readily produced in stars were it not for some 'tuning' that apparently takes place in the properties of its nucleus, which depend even more sensitively on this same number.

Even a universe as large as ours could be very boring: it could contain just black holes, or inert dark matter, and no atoms at all. Even if it had the same ingredients as ours, it could be expanding so fast that no stars or galaxies had time to form; or it could be so turbulent that all the material formed vast black holes rather than stars or galaxies, an inclement environment for life. And our universe is also special in having three spatial dimensions. In a four-dimensional world atoms would be unstable; in two dimensions, nothing complex could exist.

What (if anything) does the apparent 'tuning' mean?

If our existence depends on a seemingly special cosmic recipe, how should we react to the implicit fine-tuning? Unlike evidence for biological design, it cannot be attributed to any evolutionary adjustment – the laws of physics are 'given' and cannot adjust in symbiosis with their surroundings. There seem three lines to take: we can dismiss it as happenstance; we can acclaim it as the workings of providence; or (my preference) we can conjecture that our universe is a specially favoured domain in a still vaster multiverse.

Maybe a fundamental set of equations, which some day will be written on T-shirts, fixes all key properties of our universe uniquely. It would then be an unassailable fact that these equations permitted the immensely complex evolution that led to our emergence. But I think there would still be something to wonder about. It is not guaranteed that simple equations permit complex consequences. To take an analogy from mathematics, consider the beautiful pattern known as the Mandelbrot set. This pattern is encoded by a short algorithm, but has infinitely deep structure: tiny parts of it reveal novel intricacies however much they are magnified. In contrast, you can readily write down other algorithms, superficially similar, that yield very dull patterns. Why should the fundamental equations encode something with such potential complexity, rather than the boring or sterile universe that many recipes would lead to?

Some of course would invoke providence of design. Two centuries ago, the Cambridge theologian William Paley introduced the famous metaphor of the watch and the watchmaker – adducing the eye, the opposable thumb and so on as evidence of a benign Creator. This line of thought fell from favour, even among most theologians, in post-Darwinian times. We now

Any conceivable form of life would have to keep ever cooler, think slowly, and hibernate for ever-longer periods. But there would be time to think every thought. As Woody Allen once said, 'Eternity is very long, especially toward the end'

Astronaut Peter J. K. Wisoff is caught reflected in the helmet visor of astronaut Michael Lopez-Alegria as he took this photograph in the cargo bay of the Earth-orbiting Discovery. A cloud-covered Earth and part of the International Space Station are also mirrored in Lopez-Alegria's visor.

view any biological contrivance as the outcome of prolonged evolutionary selection. Paley was unable to draw much evidence from astronomy. But, had he been aware of it, he would surely have been impressed by the providential-seeming physics that led to galaxies, stars, planets and the 92 elements of the periodic table. Our universe evolved from a simple beginning – a Big Bang – specified by quite a short recipe. But this recipe seems rather special. A modern counterpart of Paley, the physicist and theologian John Polkinghorne, interprets our fine-tuned habitat as 'the creation of a Creator who wills that it should be so'.

If one does not believe in providential design, but still thinks the fine-tuning needs some explanation, there is another perspective – a highly speculative one. There may be many 'universes' of which ours is just one. In the others, some laws and physical constants would be different. But our universe would not be just a random one. It would belong to the unusual subset that offered a habitat conducive to the emergence of complexity and consciousness. The analogy of the watchmaker could be off the mark. Instead, the cosmos may resemble an 'off-the-shelf' clothes shop: if the shop has a large stock, we are not surprised to find one suit that fits. Likewise, if our universe is selected from a multiverse, its seemingly designed or fine-tuned features wouldn't be surprising.

Are questions about other universes part of science?

Science is an experimental or observational enterprise, and it is natural to be troubled by assertions that invoke something inherently unobservable. Some might regard the other universes as being in the province of metaphysics rather than physics. But I think they already lie within the proper purview of science. It is not absurd or meaningless to ask 'Do unobservable universes exist?', even though no quick answer is likely to be forthcoming. The question plainly cannot be settled by direct observation, but relevant evidence can be sought, which could lead to an answer. Andrei Linde, Alex Vilenkin and others have performed computer simulations depicting an 'eternal' inflationary phase where many universes sprout from separate big bangs into disjoint regions of spacetimes. Alan Guth and Lee Smolin have, from different viewpoints, suggested that a new universe could sprout inside a black hole, expanding into a new domain of space and time inaccessible to us. Guth and Edward Harrison have even conjectured that universes could be made in the laboratory, by imploding a lump of material to make a small black hole. Is our entire universe perhaps the outcome of some experiment in another universe? If so, the theological arguments from design could be resuscitated in a novel guise.

Alan Guth and Edward Harrison have even conjectured that universes could be made in the laboratory, by imploding a lump of material to make a small black hole

Smolin speculates that the laws that govern the daughter universe may bear the imprint of those prevailing in its parent universe. If that new universe were like ours, then stars, galaxies and black holes would form in it; those black holes would in turn spawn another generation of universes; and so on, perhaps ad infinitum. Lisa Randall and Raman Sundrum suggest that other universes could exist, separated from us in an extra-spatial dimension; these disjoint universes might interact gravitationally, or they might have no effect whatsoever on each other. Other universes would be separate domains of space and time. We couldn't even meaningfully say whether they existed before, after or alongside our own, because such concepts make sense only insofar as we can impose a single measure of time, ticking away in all the universes.

Parallel universes are also invoked as a solution to some of the paradoxes of quantum mechanics, in the 'many worlds' theory, first advocated in the 1950s. This concept was prefigured by the visionary science fiction writer Olaf Stapledon, as one of the more sophisticated creations of his 1937 novel *Star Maker*: 'Whenever a creature was faced with several possible courses of action, it took them all, thereby creating many...distinct histories of the cosmos. Since in every evolutionary sequence of this cosmos there were many creatures, and since each was constantly faced with many possible courses, and the combinations of all their courses were innumerable, an infinity of distinct universes exfoliated from every moment of every temporal sequence.'

None of these scenarios has been simply dreamed up out of thin air: each has a serious, albeit speculative, theoretical motivation. However, only one of them, at most, can be correct. Quite possibly none is: there are alternative theories that would lead to just one universe.

At the moment, we have an excellent framework, called the standard model, which accounts for almost all subatomic phenomena that have been observed. But the formulae of the 'standard model' involve numbers that cannot be derived from the theory and have to be found by experiment. Perhaps, in the 21st century, physicists will develop a theory that yields insight into – for instance – why there are three kinds of neutrinos, and the nature of the nuclear and electric forces. Such a theory would thereby acquire credibility. If the same theory, applied to the very beginning of our universe, were to predict many big bangs, then we would have as much reason to believe in separate universes as we now have for believing inferences from particle physics about quarks inside atoms, or from relativity theory about the unobservable interior of black holes. Such a theory must unify quantum theory, which governs the microworld, with gravity, the force that dominates the large-scale universe. For most natural phenomena, such unification is superfluous: quantum theory is crucial only in the microworld of atoms, where gravity is too weak to be significant; conversely, gravity is important only on the scale of stars and planets, where the intrinsic 'fuzziness' due to quantum effects can be ignored. But right back at the beginning, everything was squeezed so dense that quantum effects could shake the entire universe.

Einstein spent his last 30 years seeking a unified theory. With hindsight, we can see that his efforts were premature: they focused on gravity and electromagnetism, without taking cognizance of the other forces: the strong nuclear force, and the so-called weak force, important in neutrinos and radioactivity. But will such a theory – reconciling gravity with the quantum principle, and transforming our conception of space and time – be achieved in coming decades? The smart money is on superstring theory, or M-theory, in which each point in our ordinary space is conceived to be a tightly folded origami structure in six or seven extra dimensions. This now engages a large cohort of young scientists. It has not yet predicted anything new – experimental or cosmological – caused by the extra dimensions. There is still an unbridged gap between this elaborate mathematical theory and anything we can actually measure. But its proponents are convinced that it has a resounding ring of truth about it and that we should take it seriously: many are willing to bet on superstrings almost for aesthetic reasons. Edward Witten, the intellectual leader of this subject, has said: 'Good wrong ideas are extremely scarce, and good wrong ideas that even remotely rival the majesty of string theory have never been seen.'

Universal laws, or mere bylaws?

Maybe there is something uniquely self-consistent about the actual recipe for our universe, and no big bang could end up producing a different kind of universe. But a far more interesting possibility (which is certainly tenable in our present state of ignorance of the underlying laws) is that the underlying laws governing the entire multiverse may allow variety among the universes. Some of what we call 'laws of nature' may in this grander perspective be local bylaws, consistent with some overarching theory governing the ensemble, but not uniquely fixed by that theory. Consider the form of snowflakes. Their ubiquitous six-fold symmetry is a direct consequence of the properties and shape of water molecules. But snowflakes display an immense variety of patterns because each is moulded by its microenvironments: how each flake grows is sensitive to the fortuitous temperature and humidity changes during its growth. If physicists achieved a fundamental theory, it would tell us which aspects of nature were direct consequences of the bedrock theory (just as the symmetrical template of snowflakes is due to the basic structure of a water molecule), and which are (like the distinctive pattern of a particular snowflake) the outcome of accidents. The accidental features could be imprinted during the cooling that follows the Big Bang, rather as a piece of red-hot iron becomes magnetized when it cools down, but with an alignment that may depend on chance factors. The cosmological numbers in our universe, and perhaps some of the so-called constants of laboratory physics as well, could be 'environmental accidents', rather than uniquely fixed throughout the multiverse by some final theory. Some seemingly 'fine-tuned' features of our universe could then only be explained by 'anthropic' arguments, which are analogous to what any observer or experimenter does when they allow for selection effects in their measurements: if there are many universes, most of which are not habitable, we should not be surprised to find ourselves in one of the habitable ones!

We may one day have a convincing theory for our Big Bang that tells us whether a multiverse exists, and (if so) whether some so-called laws of nature are just parochial bylaws in our cosmic patch. What we have traditionally called 'the universe' may be the outcome of one big bang among many, just as our solar system is merely one of many planetary systems in the galaxy. The forces governing our universe, and the mix of atoms, radiation and other particles that make it up, may be 'cosmic environmental accidents', rather than uniquely defined by the underlying theory. The quest for exact formulae for what we normally call the constants of nature may consequently be as vain and misguided as was Kepler's quest for the exact numerology of planetary orbits. Other universes will then become part of scientific discourse, just as 'other worlds' have been for centuries. Nonetheless – and here physicists should gladly concede to the philosophers – any understanding of why anything exists, of why there is a universe (or multiverse) rather than nothing, remains in the realm of metaphysics.

Bitter medicine

Sarah Boseley

Until an international outcry forced them to withdraw their action, the drug companies tried to block changes in South Africa's patent law. Millions are threatened by Aids, but have lives been saved?

In March 2001, 39 pharmaceutical companies brought what they considered was a routine court case against the government of South Africa. The initiative was taken on their behalf by the Pharmaceutical Manufacturers Association of South Africa, as the local representative body, but their action had the approval of the lawyers and chief executives of the multinational drug giants in the UK, USA and Europe. At stake, as they saw it, was the vital principle of the inviolability of drug patents.

For the previous three years, the drug giants, aided by their friends in the US administration, had been locked in argument with Nelson Mandela's and then Thabo Mbeki's government over South Africa's right to obtain cheap medicines from abroad. Under the World Trade Organization's TRIPS (trade-related intellectual property rights) agreement, poor countries facing a health crisis are permitted to disregard patents and make their own cheap, generic copies of drugs they need, or bring them in from elsewhere. But countries are obliged to pass enabling legislation first.

South Africa was not the first nation to fall foul of the multitudes of patent lawyers retained by the drug companies to scrutinize such laws. They took issue with the detail of the South African bill. It was too sweeping, they protested. It would allow the wholesale import of cheap medicines. It would set a precedent for the rest of Africa and poke a hole in the dyke of the patent protection system. If this sort of thing were allowed, they argued, drug companies would go out of business, unable to make the huge profits they need to recoup the millions of dollars they spend on research and development. A date for the hearing against Mandela and his government was set in Pretoria's High Court.

While the lawyers were preparing their briefs, an epidemic that has been compared to Europe's Black Death was laying waste ever more people

> Not a single South African has been put on a course of antiretroviral drugs as a result of this historic climb-down by the drug companies. The government never intended to buy ARVs

in sub-Saharan Africa. For a long time, African nations denied it was happening, but first Uganda, then gradually others, admitted that they were suffering the depredations of HIV/Aids.

In July 2000, the International Aids Conference was held for the first time in a developing country badly hit by the disease. The event in Durban, South Africa, was less a scientific meeting than a highly charged emotional rally. Its theme was 'Breaking the Silence'. Taboo-breakers like the tiny HIV-positive boy Nkosi Johnson and the High Court Judge Edmund Cameron made headlines around the world. It proved a watershed. In South Africa, the clamour of grassroots activists began to be heard, demanding medical treatment for the millions for whom an HIV diagnosis was a death sentence.

But Aids drugs cost around $10,000 a year per patient – way beyond South Africa's reach – and were protected by patents. Thirty-nine pharmaceutical giants were prosecuting the South African government to make sure things stayed that way.

There was an international public outcry at the spectacle of multinationals, some with sales figures higher than the GDP of certain developing countries, preventing access to Aids drugs. This eventually forced a rethink in the drug-company boardrooms. An order by the South African judge that they should disclose details of their pricing policies may also have been a factor. At the end of a six-week adjournment, the 39 drug companies threw in the towel and dropped the case. Aids campaigners were ecstatic.

Not a single South African has been put on a course of antiretroviral drugs (ARVs) as a result of this historic climb-down by the drug companies. The government never intended to buy ARVs, although it has said that it wants to import cheap drugs for the opportunistic infections that kill those whose immune systems are damaged by the HIV virus.

Thabo Mbeki, the South African president, risks making history as the leader who failed his people by turning his back on science. Stubbornly refusing to believe that HIV is the cause of Aids, he prefers the theories of the maverick American scientist Peter Duesberg, whose insistence that Aids was not a viral infection was largely discredited ten years ago in the USA. In the face of considerable international pressure, Mbeki maintains that Aids is caused by poverty, undermining the efforts in his country to promote safe sex and pushing treatment for the disease way down the agenda.

But the High Court hubris of the drug companies has had enormous impact. It sent a signal to poor countries that patents were not necessarily unassailable and that the WTO agreement could be interpreted in their favour. A coalition of African nations succeeded in obtaining a declaration at the WTO ministerial meeting in Doha in November 2001 that the trade agreement on patents for medicines should not get in the way of public health.

It has also sent drug prices spiralling downwards, thanks to generic companies in India – not yet bound by TRIPS – which offered rock-bottom prices for their own copycat Aids drugs, forcing the drug giants in turn into giving substantial discounts. Prices reached a low of under $300 per patient per year – still far too much for most people in most poor countries, but encouraging some governments to offer treatment, albeit only to a minority.

Over 2 million Africans die every year of HIV/Aids. In South Africa, around a quarter of young adults are infected. According to the latest World Health Organization figures, nearly 25 million people around the world have died since the epidemic began, while 40 million are living with HIV. The chances of medication for the millions with HIV in sub-Saharan Africa and in Asia, where it is now recognized an HIV explosion is taking place, are slim. Vast areas have neither hospitals nor sufficient drugs for killer diseases like diarrhoea, which ought to be easily cured. But the court case, followed by a UN special session and the launch by UN General Secretary Kofi Annan of a global fund for Aids, malaria and tuberculosis, has at least begun to concentrate minds on the scale of the problem.

Something extra

Tim Radford

ANDi, the world's first transgenic monkey, was announced with a glow of triumph in January. The little baby rhesus monkey carried a gene that coded for green fluorescent protein, which had been isolated from a jellyfish. A team at the Oregon Health Sciences University in the US injected a genetically modified virus into unfertilized eggs and then added sperm. There were 224 eggs treated, 20 embryo transfers, but only five pregnancies. A pair of twins miscarried, but three healthy monkeys were born. Of these, only ANDi – backwards for inserted DNA – was confirmed as carrying the jellyfish gene. The dead twins also carried the gene, and unlike ANDi, their hair follicles and toenails glowed under fluorescent light.

Transgenic mice are now routine laboratory tools, and there have been transgenic pigs, sheep and cattle, but ANDi is a primate, and represented a step on the way to understanding human illnesses. 'In this way, we hope to bridge the gap between transgenic mice and humans. We could get better answers from fewer animals, while accelerating the discovery of cures through molecular medicine,' said ANDi's creator Gerald Schatten. A baby gaur called Noah seemed, for a few hours, to demonstrate a role for cloning to anxious conservationists. A gaur is a wild ox native to Southeast Asia. It is also an endangered species. Scientists at Advanced Cell Technology in Massachusetts fused skin cells from a male gaur with cow eggs, and transferred 44 embryos into 32 surrogate mother cows to produce eight pregnancies, five miscarriages and two foetuses which were removed for tissue examination.

Noah was born by caesarean section on 8 January 2001, was pronounced healthy – and died two days later of diarrhoea. During 2001, an Italian fertility specialist and two other groups with less expertise claimed to be intent on cloning a human being, to the alarm of those researchers who had successfully cloned mice, or farm animals, and were more aware of the difficulties.

The ring cycle

Mike Baillie

Imagine a wooden register of the seasons for every year

In the jungle of scientific debate, you cannot always see the wood for the trees. But in climate change, the wood itself sometimes holds the key. Imagine an annual register of a year's sunshine and rainfall and frost, kept up to date with perfect accuracy almost everywhere south of the tundra and north of the tropics, and available for inspection not just at any time in life but, quite often, for centuries after death. The register is, of course, the annual growth rings in trees. Match the rings from young trees with those from old forest giants and you have a centuries-long measure of the march of the seasons. Match the rings from old trees with old cathedral rafters and you have a still longer chronology – and a science called dendrochronology.

Dendrochronologists, scientists who study the growth rings of trees, have successfully constructed long tree-ring records by overlapping the patterns of wide and narrow rings in successively older timber specimens. There are now a dozen or so chronologies in the world that date back more than 5,000 years. These records, normally constructed in a restricted area, using a single species of tree, are year-by-year records of how the trees reacted to their growth conditions – an environmental history from the trees' point of view.

Why an article about the science of dendrochronology in 2001? In fact, one could write an article about the latest advances in dendrochronology any year. That is the nature of developing fields of study – there is always something going on. The 1990s saw a lot of interest in issues such as drought reconstruction and volcanoes. This is because, in some areas, trees are highly sensitive to drought; and because archaeologists need to date certain volcanoes, such as Santorini in the Aegean – and it helps that many pine species are sensitive to the cooling effects brought on by volcanic dust veils.

Because tree-ring chronologies are constructed on a regional basis, there has, in the past, been a

Match the rings from young trees with those from old forest giants and you have a centuries-long measure of the march of the seasons. Match the rings from old trees with old cathedral rafters and you have a still longer chronology – and a science called dendrochronology

tendency for dendrochronologists to think local. However, the success of dendrochronology as an international research topic means that there are now quite a lot of chronologies available for study. As the chronologies are dated absolutely, it is possible to compare the records from different areas year by year. Recently, an analysis of 383 modern chronologies, drawn from a vast area across Europe, northern Eurasia and North America was published. The authors, Keith Briffa and colleagues, observed that the maximum late-wood density of the growth rings in each year was related to the temperature in the growing season. Their analysis spanned 600 years, back to AD 1400, and presented a summer temperature record reconstructed from the huge grid of precisely dated ring densities. What they noticed was that the years of really low density – the cool summers – were directly associated with large explosive eruptions, as known from historical sources and from dated layers of acid in the Greenland ice record. Greenland ice is kilometres thick and is made up of the compressed snowfall of tens of thousands of years, so the ice record can be read in almost the same way as tree-rings. I shall use this study as an example of what else tree-rings can tell us.

The study provides a year-by-year estimate of temperature together with the dates of some major volcanoes. It is a nice clean story – volcanoes load the atmosphere with dust and aerosol and reflect back sunlight, cooling the earth's surface. This cooling leads to variations in the density of growth rings in northern conifers. Because there are a lot of other records, it is possible to test the findings from the pine density record.

We can, for example, look at what European oak was doing across the same 600-year period. Was oak responding in the same way as the conifers? The 'oak chronology' is the mean of eight regional oak chronologies across a strip of land from Ireland to Poland; it represents how,

Clumber Park, Worksop

on average, hundreds of millions of oaks grew. What we see from this comparison is that the oaks clearly do respond to the volcanoes in some cases (in 1601, 1740 and 1816, for instance), but nothing like so clearly in others. Immediately it becomes apparent that the conifers tell only part of the story. There are many downturns in oak growth, and only a few are related to the conifer record. The oaks were quite capable of being more stressed in years where the conifers were not affected. The point of this, however, is not to argue about the quality of global cooling – the point is to show what dendrochronology can do.

Take the case of 1816, called the 'year without a summer' because of the terrible unseasonable cold and the crop failures that ensued. It has long been known that the primary cause of the cooling was the massive eruption of Tambora, east of Java, in 1815. However, there was a lot going on in the run-up to 1815. Bald cypress trees in Tennessee show a major growth anomaly, with rings up to 400 per cent wider than normal, in the years following a huge earthquake in 1811–12 in Eastern America. But there is a volcanic acid layer in several Greenland and Antarctic ice cores in 1809–10, as well as in 1815–16. So here we have a combination of a highly unusual quake in an area of the USA not normally affected by earthquakes, and at least two volcanic eruptions, including Tambora, which is widely regarded as the largest in the last 10,000 years. According to Briffa, the decade 1810–20 was the coldest in the last millennium, so we begin to see a combination of three unusual elements in less than a decade – exceptional earthquake, exceptional volcanic eruption and exceptional cold. Given that the defeat of Napoleon's invasion of Russia in 1812 was famously attributed to 'General Winter', one wonders whether a natural series of events actually helped to change the course of modern history.

Obviously, the case of 1816 is relatively recent and well documented. However, dendrochronology allows us to investigate the effects of such events geographically, indeed globally. We can interrogate the trees in areas where there is no historical or instrumental record. Further back in time, dendrochronology is almost the only way to reconstruct abrupt environmental events and perhaps throw light on far darker moments in human history. Were there just political forces at work in the Dark Ages, or did violent natural events also take a hand, tipping the balance by darkening the skies and lowering the temperature? The trees were there too, and kept a record. The wood hewn from them and preserved through the centuries is slowly beginning to yield at least circumstantial evidence that could support some of the stories – think of the Arthurian wasteland, or the plagues of Egypt – so far told only in enigmatic artefacts, or in legends, epics and religious chronicles.

Death from underfoot

Ben Wisner

2001 saw two catastrophic earthquakes

El Salvador, 13 January 2001. It was 10.33 on a Saturday morning and shoppers were busy in the streets of Santa Tecla, one of the suburbs that snake out in narrow valleys among volcano slopes and outlier ridges from the old heart of the capital, San Salvador. On farms and in rural towns and villages, people were busy making their living producing coffee and bananas for export, as well as rearing livestock and growing maize, beans and rice for local sale.

Fifteen seconds before, some 40 kilometres deep in the strata below the Pacific Ocean, rock that was being forced still deeper had suddenly cracked under the stress. A burst of energy exploded from this crack in the Cocos Plate, one of six massive pieces of the lithosphere that float upon Earth's hot mantle, jostling in slow motion at this point on our planet's 'ring of fire'. The energy released was equivalent to more than a thousand million kilograms of high explosive. It rushed out in all directions from the fault in a series of four different kinds of waves, travelling at various speeds and amplitudes. First the earth gave a sudden jolt, like the shunting of a long train as it begins to move. Then it began to undulate.

Above the suburb of Santa Tecla, the ground acceleration from the 7.6 magnitude earthquake loosened the moist soil, derived from young rocks made of wind-blown volcanic ash. On the steep slope, trees had already been cut down in preparation for a set of new luxury homes. The hillside crumbled and slid down on to a residential neighbourhood called Las Colinas. Four hundred houses were buried, and 700 people died. There were thousands of aftershocks over the next month. Schoolchildren who were in buildings that had been damaged in January died when they finally collapsed after another earthquake in February.

Other landslides blocked the Pan American Highway and many other roads. Elsewhere in the country, unreinforced brick and adobe structures

Schoolchildren who were in buildings that had been damaged in January died when they finally collapsed after another earthquake in February

collapsed: homes, granaries, warehouses, churches, clinics and schools. In all, the two earthquakes and thousands of aftershocks left 1,159 people dead and 8,122 injured, and destroyed more than 150,000 homes. An additional 185,000 homes were damaged. Infrastructure was also heavily affected, with damage to 1,566 schools, 23 hospitals and another 121 health care units – roughly 40 per cent of the nation's hospital capacity and 30 per cent of its schools. The total economic loss was estimated at $1.255 billion. For a small country of 6 million people, a society struggling to reorganize itself following a long civil war, and with a weak economy, El Salvador's earthquake losses were significant. The economic losses amounted to 10 per cent of GNP and one half of the annual national government budget.

Less than two weeks later, on Republic Day, 26 January 2001, India suffered its worst earthquake in 50 years. The energy release was one hundredth of that caused by the intraplate faulting off the southern coast of El Salvador. Yet this 6.9 Richter-scale earthquake still released energy equivalent to nine Hiroshima-size atomic bombs. The epicentre was shallow (24km) and more or less directly under a zone of impoverished towns and drought-stricken villages in Gujarat, not far from the border with Pakistan.

It was 8.47 in the morning. Without warning, the ground heaved and buildings shook for two minutes, subjected to accelerations of half the force of gravity. This is comparable to the stresses created in El Salvador by the larger, but deeper and more distant earthquake, and was more than enough to tear poorly constructed buildings apart. Hundreds of schoolchildren and their teachers parading in a narrow street as part of Republic Day celebrations were killed instantly. The towns of Bhuj, Bachau, Rajkot and Anjar were devastated.

Even in India's large cities it is estimated that only one in ten structures is built according to the country's seismic code. Unreinforced buildings in these small towns crumbled. In Gujarat's largest city, Ahmedabad, high-rise apartment buildings that had been constructed with little attention to building standards also collapsed, killing 750. In all, at least 20,000 people died. Many were injured, and 600,000 were left homeless. Nearly a thousand schools and many health facilities were destroyed.

Action at national and local level is required to save lives in earthquakes. Comprehensive planning and strong institutions, both anchored firmly in an enforceable system of laws, are clearly needed. Local citizen activism is also vital. Residents and city officials in Santa Tecla had been in court to stop the deforestation and construction of homes on the slope above the ill-fated Las Colinas neighbourhood. The whole of the Balsamo range of hills was known to be unstable, yet the court had found in favour of the developer. How could this happen? In India, scientists and non-governmental organizations have been trying to improve central and state government planning for disasters for years. So why are building codes still not enforced there? One lesson that seems to have been learned locally in El Salvador and India is that far more political pressure must be exerted to force reforms. A law that would set in place an accountable and comprehensive system for disaster prevention has languished in El Salvador's National Assembly since Hurricane Mitch swept through Central America in 1998. A broad coalition of citizens' groups is now lobbying for this legislation to be passed.

In both El Salvador and India, local NGOs are trying to direct aid toward projects that will make communities safer in future extreme events. A wide range of groups have been involved, including those with skills in popular education, environmental groups such as UNES in El Salvador, and women's groups such as the Self-employed Women's Association in India. In a unique combination, a group of popular educators based in San Salvador called Equipo Maiz collaborated with the National University, the environmental group UNES and foreign donors to produce, in comic-book format, a manual for constructing reinforced adobe houses. The coalition has also built a number of prototypes.

Earthquakes on this scale will always occur. But to prevent injury and death, and to reduce economic loss, social action is needed: comprehensive and effective laws and institutions from above, and from below, awareness and efforts at self-protection on the part of ordinary citizens.

The whole of the Balsamo range of hills was known to be unstable, yet the court had found in favour of the developer. How could this happen?

The sixth wave

Tim Radford

Biodiversity is a six-syllable word that tends not to be on everybody's lips. During 2001, researchers added a number of species to the catalogue of life and almost certainly said goodbye, without knowing it, to a much larger number. Researchers from Kew in London, St Petersburg in Russia and the Missouri Botanical Gardens in the US identified a new species of conifer in the karst landscape of northern Vietnam, and named it *Xanthocyparis vietnamensis*. Actually, they had gone there looking for orchids. A US-Puerto Rican group claimed to have found the world's smallest gecko. *Sphaerodactylus ariasae* lives in the Dominican Republic and is just three quarters of an inch long. 'We did not even know the species existed, although the area has been studied by biologists for several hundred years,' said Blair Hedges of Pennsylvania State University.

Lord May, president of the Royal Society, pointed out in November that biologists did not know to within a factor of 10 how many species the planet was home to – estimates ranged from 3 million to 100 million. But species were almost certainly disappearing at a rate between 100 and 1,000 times faster than the background rate of evolution. 'We are standing on the breaking tip of the sixth great wave of extinction in the history of life on Earth,' he said at a lecture at the Natural History Museum in London. The United Nations Environment Programme had already launched a campaign to save the great apes: chimpanzees, gorillas and orang-utans are all threatened by hunters, poachers and habitat destruction. And scientists in China had gloomy news for those who believed that wildlife reserves must necessarily provide hope for wildlife. In 1975, the Chinese established the Wolong reserve as a flagship conservation project for the giant panda, itself an emblem of threatened wildlife. In April 2001, they reported in *Science* that they had studied two decades of satellite images. To their distress, after 25 years, the reserve contained less forest cover suitable for panda habitat than it did when it was first established. 'If biodiversity cannot be protected in protected areas, where can we protect biodiversity?' asked Jianguo Liu of Michigan State University.

Space odysseys

Paul Murdin

2001: the year that Mir fell to Earth, and a tourist checked into the International Space Station

After 15 years in space, the Mir space station was taken out of orbit on 22 March 2001. Its 140 tonnes plunged into the atmosphere in a fiery blaze over Fiji, splashing debris into the southern Pacific Ocean. Rosaviakosmos, the Russian space agency, had attempted to prolong the lifetime of Mir beyond its natural limits by offering it as a hotel for tourists. Its hopes were desperate and futile. Launched in 1986, Mir was the culmination of the Russian effort to maintain human presence in space. From the flight of Yuri Gagarin to the demise of Mir, Russia was supreme in manned space flight. Mir poignantly carried into the sea the symbolic status of Russia as a global power.

At the heart of Mir (it means 'peace', and also 'commune', 'village' or 'world') was a module where the cosmonauts lived and worked in a shirtsleeve environment. A docking port permanently held a Soyuz ('Union') spacecraft, both as an extension and as a potential lifeboat. Progress and Soyuz supply-ships, and the US Space Shuttle, docked at Mir's other ports and installed laboratory modules. Mir was the model for the International Space Station.

During the lifetime of Mir, increasingly fewer supply-ships visited. This was partly for financial reasons, but also Mir learnt how to be more self-sufficient. It more efficiently recycled materials like water so that reserves did not have to be topped up so frequently. The last automatic docking to the empty space station took up the rocket fuel that brought Mir down. Mir was extremely successful. It had, however, been configured and reconfigured by cosmonauts, determined more to achieve their urgent goals than to preserve future capability. Live TV transmissions showed the results of the Russian 'make do and mend' philosophy. Cables between the modules had been laid through the hatchways. When the Spektr ('Spectrum') module began to leak air after a docking collision in July 1997, the cosmonauts had to sever cables before the isolating hatchway could be closed. Thanks to the cosmonauts' daring, Mir's occupation record was maintained, but in the final months, Mir was operating at the limits of its safety.

The occupation record of Mir in its prime was impressive. The almost continuously manned space station hosted two to three cosmonauts (on occasion up to six for short periods). Eighty astronauts spent a total of about 25 years in Mir, a dozen for six months at a time, three of them for a year or more. Like many adventures, the main outcome of Mir's odyssey was really the adventure itself. Certainly the lessons of the Mir experience – human necessities, limits of safety, ground support, logistics of supply, how to recycle – were sought by NASA for use in the International Space Station. The search for scientific knowledge was an oversold promise. What final verdict will history cast on Mir? Was it a grand but

From the flight of Yuri Gagarin to the demise of Mir, Russia was supreme in manned space flight. Mir poignantly carried into the sea the symbolic status of Russia as a global power

expensive nationalistic folly, or the precursor of mankind's permanent colonization of space?

The International Space Station

Mir's nominal seven-year lifespan had been doubled, an achievement in its own right. This was to overlap with the construction of the International Space Station (ISS). The 450-tonne, 1,000-cubic-metre ISS has the capacity of three American houses. It will cost NASA $17 billion up to 2002, including about $1 billion paid to Rosaviakosmos for the contributions that had been arranged but which, in its current economic climate, Russia cannot afford. The continuing bill is some $2 billion per year. Canada, Japan, 11 of the 15 European Space Agency (ESA) member states and some further countries make additional contributions.

The first modules of the ISS were launched into independent orbits in 1998–9 and assembled in 2000 by astronauts from Space Shuttle and Proton rocket flights. Three expeditionary flights (October 2000, March 2001 and July 2001) have begun the permanent occupation of the ISS, bringing it to functioning status. Against NASA's humourless objections to bringing a passenger to the ISS at such a risky early stage, Rosaviakosmos controversially hosted a paying guest at the ISS in April 2001. Space tourist Dennis Tito paid a reported $20 million for his package holiday. He spent it at the station, floating cheerfully in weightlessness, watched by his admiring wife on TV. America watched too, torn between envy and derision.

As it grows in size, the ISS is becoming even more readily visible (about as bright as the planet Venus) as it orbits overhead after sunset and before sunrise. Visibility predictions for anywhere in the world are available on the World Wide Web at http://spaceflight.nasa.gov. When the ISS is being approached by the Space Shuttle on a visit, it is exciting and inspiring to watch the two man-made stars pass overhead as a pair.

The ISS's first science experiments got under way in the second half of 2001. Until 2001, the ambitious plan was to complete and operate the International Space Station with a total of more than 40 space flights, at almost monthly frequency. Cost overruns forced a rethink, and provoked the resignation of NASA's otherwise highly successful chief administrator, Dan Goldin, in October 2001. The US Congress's refusal to fund the overruns has reduced planned flight frequency, crew numbers and research time. The plan now includes 'core' and internationally agreed items. NASA will put forward optional add-ons in 2002 to a sceptical US Congress, asking for an increased budget. The European Space Agency is helping make up the gap with further contributions, principally from Italy and an increasingly reluctant France and Germany.

Magic carpet

In January 2001, China launched and safely landed a Shenzhou spacecraft. This was Shenzhou's second test flight, expected to lead to manned flights. The spacecraft orbited Earth 108 times, containing a Noah's ark of small animals, from monkeys to snails. Shenzhou ('Magic Vessel') is like the Russian Soyuz and American Apollo capsules. It will carry two Chinese astronauts. China intends more test flights but is coy about their timing and the intended first manned flight. A developing space power, China has launched nearly 300 satellites. In November 2000, it published ambitious plans for a major space programme, including its own space station in 2015 and a man on the moon.

2001: a space odyssey delayed

2001 was a year given red-letter status for space enthusiasts by Stanley Kubrick and Arthur C. Clarke's film *2001: A Space Odyssey*. If the film's projections had been realized, space travel would now be almost routine. It envisaged space colonized by giant hotel-like rotating space stations. One of the lessons from Mir was that it was not necessary to produce artificial gravity, which is why Mir and ISS look nothing like the space station in the film. Space visionary Gerard O'Neill foresaw a similar future, much like Ridley Scott's film *Blade Runner*, with extensive manned occupation of space, in orbit and in lunar mining colonies. In 1975, O'Neill predicted that a quarter of a million people would be stationed off-world by 2005. The actual number in 2001 averaged about three.

The vision is slowly coming to reality. The reasons are the cost and the risk. Inanimate payloads cost about $10,000 a kilogram to orbit around Earth, people much more. Mr Tito's $20 million return ticket to the ISS was realistically priced, and there are few customers in the tourist market

The International Space Station photographed by one of the astronauts on Atlantis. The crew have been up there since November 2000.

Some planned missions crept over budget or looked too risky. In 2001, cancellation was threatened or implemented for several missions, for example the first mission to Pluto

as rich as he was. O'Neill's forecasts were based on a predicted $3,000 cost for a manned space flight. When the cost reduces towards this figure, space travel may take place on the scale of business class air travel.

The risk of a rocket launch is not negligible. The design aims of ESA's new Ariane 5 rocket included unprecedented safety margins. However, it has delivered its payload into the wrong orbit (or not at all) three times in its five launches up to 2001. In July, it put ESA's experimental telecommunications satellite, Artemis, into an orbit that was too low. Artemis' UK-built ion engines (intended for small station-keeping adjustments) will take a year to nudge it valiantly into the right orbit, like a sailor using oars to get to the destination after the outboard motor has failed.

The Space Shuttle has just about achieved its designed risk level (one serious failure – Challenger – in 100 launches, with 150 the design target). Even so, the risk is about 1 per cent. Adventurers and test pilots accept this level, but not most tourists. The mass exploitation of space travel depends on the development of a safe, reusable space plane. The difficulty of this was illustrated in 2001 by NASA's cancellation of its X33 and X34 space plane projects due to insuperable difficulties with the engine technology.

The final frontier

Robotic spacecraft are proxy explorers for the human race. There are an increasing number of them. The early missions, like Voyager, showed us eight of the solar system's nine major planets (no spacecraft has visited Pluto) and many of its 60 moons as individual worlds. Some of these are like our own moon, pitted with craters from the impacts of meteors, asteroids and comets. The surfaces of these worlds are a palimpsest from which astronomers can read the massive celestial bombardment of the first millions of years of the solar system. * Other worlds are more Earth-like, but show the effects of frozen atmospheric devel-

* The residual bombardment continues, with Earth last struck by an asteroid or comet 50–100 metres in diameter in 1908. Following political activity by the pressure group Spaceguard, the UK Space Minister, Lord Sainsbury, set up a British task force to study the issue. The task force reported in 2000 that the risk from impacts by similar so-called Near Earth Objects (NEOs) was real. As is the way of governments, however, its recommendations are being set in place too slowly over the next few years to satisfy Spaceguard. They are likely to include increased telescope power to find, track and characterize NEOs.

opment (Titan), a runaway greenhouse effect (Venus), massive global climate change (Mars) and global volcanism (Io). These are a few of the horrors that the evolution of Earth has avoided.

The evolutionary links between these many different kinds of planets – examples of what our world might have been and how it might develop – are becoming clearer. With this perspective, and from direct observation of Earth through so-called 'remote sensing' from space, scientists have a growing understanding of how our world works. One day, we may have to use this information to manage the future of the planet.

At a meeting of Space Ministers in Edinburgh in November 2001, ESA adopted the Aurora Programme, which will extend ESA's one-at-a-time scientific planetary missions and look towards manned space flight beyond the ISS. By 2020, Europe will have enough knowledge to decide whether to put a European on Mars.

NASA's comprehensive planetary programme sets out to explore Mars with a fleet of 'remote sensing' spacecraft. Other spacecraft will land 'rovers' on to Mars to roam and investigate chosen sites of special interest. The plan will culminate with a spacecraft that will return from Mars with a sample of soil. The return will be to quarantine in the International Space Station, as a planetary protection measure.

The loss of NASA spacecraft at Mars, particularly the fiasco of the loss of Mars Polar Lander in 1999 through an engineering muddle between metric and non-metric units, damaged confidence in the programme. Some planned missions crept over budget or looked too risky. In 2001, cancellation was threatened or implemented for several missions, for example the first mission to Pluto.

On the other hand, NASA's many successes shone out in 2001. The Galileo probe, launched in 1995, is still in orbit in the Jupiter system. The mission is in its third extension. The mission's main objectives have been safely achieved, including a landmark discovery of the under-ice ocean of Europa. Galileo's controllers are getting more daring. In 2002, they plan the probe's closest approach to Jupiter's moon, Io, to study its active volcanoes from close range. If Galileo survives this, it will subsequently over-fly the moon, Almathea, by 500km and pass through Jupiter's rings of orbiting rocky meteors (less populated versions of Saturn's rings). If Galileo's luck still holds, it will venture as close as possible to investigate Jupiter. It will finally be disposed of by a death plunge into Jupiter's atmosphere in 2003.

ESA's SMART-1 mission is being prepared to explore the Moon late in 2002. Its exact launch date is uncertain because, like an impecunious backpacker, it will stand by to ride in the next suitable Ariane 5 launch with a full-fare business payload and spare capacity.

NASA's and ESA's Cassini and Huygens spacecraft are on their way to Saturn. Launched

ESA is scheduled to be the next space power on Mars, at Christmas 2003, with the Beagle lander. Beagle, currently under construction in the UK, will carry equipment to analyse Martian soil to to determine the planet's history

in 1998, they have executed a complex manoeuvre to gain speed at Venus, Earth and Jupiter, and will arrive at Saturn on July 2004. The Huygens daughter probe will parachute from Cassini on to Saturn's moon Titan, relaying data to Earth via the mothercraft. In-flight tests in 2001, however, showed that radio communication between Cassini and Huygens was not working properly. A design blunder had overlooked radio frequency changes as Huygens drops on to Saturn, which would have left the two spacecraft unable to talk to each other. Controllers have replanned the trajectories and installed specially designed software to iron out the problem.

NASA's Mars Global Surveyor has been in orbit over Mars since 1997, discovering geological features in close-up pictures and surveying the whole planet to an accuracy of 30 metres. In the past, it has amassed evidence about the effects of water on Mars' surface. NASA's Mars Odyssey arrived safely at Mars in October 2001. It will analyse Mars' atmosphere and its radiation environment.

ESA's and NASA's programmes for the exploration of Mars aim to discover life on the planet, living or fossilized. As shown first by the Viking landers in the 1970s, Mars changed from a wet and warm climate to its current fierce, dry, desert climate. If life developed there during the more benign period, it may have survived by moving into suitable niches (near geothermal springs for example). Its discovery would be of profound philosophical interest and high scientific value. It will be both similar to and different from terrestrial life, and the 'compare and contrast' exercise will reveal aspects of terrestrial life that we didn't know were important.

ESA is scheduled to be the next space power on Mars, at Christmas 2003, with the Beagle lander, part of ESA's Mars Express mission. Beagle, currently under construction in the UK, will carry equipment to analyse Martian soil, to determine the planet's history. In particular, Beagle scientists hope to discover whether any of the soil shows biological origins. Beagle's scientific equipment has been allocated minute spacecraft resources – the same mass as a couple of bags of sugar, the same volume as a shoebox, and the power of a penlight torch. Its on-site analysis of Martian material will have to be more focused than laboratory analysis of a returned sample, but control of terrestrial contamination is surer. Beagle was on track in 2001 for its fast-approaching delivery date.

Voyages of the imagination

Astronomical missions enable scientists to voyage to other parts of the universe in their imaginations. Despite NASA Administrator Dan Goldin's attempts to wean astronomers off the Hubble Space Telescope and move on to the Next Generation Space Telescope (NGST), HST continues to deliver revolutionary observational results. NASA's Chandra X-ray telescope continued operation in 2001, and provided indications of the existence of a new kind of black hole. Stars that explode as supernovae make solar-mass black holes, and galaxies make black holes millions or billions of times more massive. Chandra has discovered X-ray stars that lie between the extremes. The question is what made these black holes, and how. ESA's X-ray Multi-Mirror telescope, the Newton Observatory, is on a similar track. ESA's Integral satellite will carry a gamma ray telescope at its launch on a Russian Proton rocket scheduled for October 2002. It will view gamma rays from very energetic phenomena, including gamma ray bursters, which are possibly neutron stars in a binary star system merging to form a black hole.

NASA's Microwave Anisotropy Probe, MAP, was launched in June 2001, due to arrive on its distant observing station at the end of the year. It will survey the cosmic microwave background radiation, the remnants of the fireball produced during the Big Bang. Its structures are the first to appear in the universe. They grew into the largest objects in the universe today, like the 'Great Wall', a sheet of galaxies and galaxy clusters some 500 million light years in extent. ESA's Planck satellite is scheduled for a 2007 launch, with a similar purpose to MAP. European scientists point out that Planck will be more accurate than MAP, although six years behind in the race. Nevertheless, MAP will undoubtedly steal some of Planck's thunder in this competitive and fast-moving field.

Some of the space flights of 2001 were almost short sprints. Others took, or are taking, a significant fraction of a working lifetime – 'odysseys' indeed. MAP and Planck are at the extreme: virtual odysseys, on the longest journey there is, 14 billion light years back to the Big Bang.

Stanley and Arthur

David E. H. Jones

Three decades ago, Stanley Kubrick and Arthur C. Clarke looked at the future. Last year, it arrived.

Stanley Kubrick's film *2001: A Space Odyssey*, based on a screenplay written by Arthur C. Clarke, was made in 1968, before any real human experience of space. It gives us the chance to see 2001 as it was imagined then, and to compare it with the world we know now. Like any film, it was meant as entertainment; even so, Clarke uses true science where he can. In my favourite scene, for example, the astronaut David Bowman has to get back into the spaceship Discovery in the vacuum of space, without his space-helmet. He fires the explosive bolts of his small space-pod, dives into the emergency airlock, and tumbles around to close the door. Only when he succeeds is there any noise at all. The whole scene takes place in utter silence, until the triumphant roar of entering air announces his success. This is exactly and dramatically right. There's no sound in the vacuum of space. But could it happen? NASA may know how low a gas-pressure human beings can stand, and for how long. But I don't.

Two assumptions Clarke made in 1968 are still valid in 2001. First, a journey to Jupiter will take a very long time: in the film, most of the crew hibernate in refrigerated suspended animation. This makes good physical sense. Second, the best rocket fuel will be very light and make the rocket go very fast. Hydrogen is the lightest element; we burn it in oxygen to make the hydrogen-oxygen rocket, the basis of the Shuttle (among other rockets). In 1968, it was hoped that by 2001 nuclear reactors would heat the hydrogen. This is how the spaceship Discovery works. But even with cheap nuclear energy to play with, fuel economy is crucial. Discovery expels hot hydrogen for a few minutes only, at the beginning and end of its long voyage. The rest of the time it is coasting like a planet, taking many months to get round its fraction of an orbit. By heating more hydrogen, it could go faster and get there sooner; but in 1968 (as in 2001), we could not afford such profligacy. Kubrick omits hydrogen tankage from the visual design of Discovery, and Clarke uses the long voyage-time in his plot. Correctly, Discovery was clearly built in, and for, space: it could not withstand earthly gravity or air drag, nor lift itself against them.

The science and technology of space is well shown. The utter blackness of space itself, with the stars as pinpoints of light, is now so familiar that we forget how bold it seemed in 1968; the sharp, clear images of solar space-light contrast strongly with the diffuse illumination we know on Earth. Indeed, my main quarrel with the science is the scene in which a space 'bus' leaves the lunar base at Clavius to travel to Tycho, where a strange black monolith has been unearthed. The 'bus' travels in a straight line. This is a mighty waste of fuel. In reality it would be accelerated by one brief thrust, would travel in an elliptical arc, and would be decelerated by reverse thrust at the far end. I suspect that Clarke was overruled on these shots.

In his predictions of electronics, Clarke is on surer ground. In 1968, he anticipated how computers would communicate with their masters, not by printer as in 1968, but by visual display. The lovely scenes in which the space-liner matches its rotation to that of the wheel-like space station in order to dock with it show this well. Clarke's communication systems and video-phones are now familiar to us, whereas they were revolutionary in 1968. The line-scan and raster of his Bell video-telephone come on, and in the right order. The displayed price of a call ($1.70 for a minute of TV connection from space) also reflects the fall in long-distance phone costs. Even the parabolic transmission aerials on Discovery make good technical sense. We still lack intelligent humanoid computers like HAL, though: we have fast stupid ones instead.

In contrast to his physical science, Clarke's biological predictions make less sense. The plot assumes that unseen aliens started certain apes on their long evolution to humanity. Our evolution still puzzles us in 2001. In a mere few million

In his predictions of electronics, Clarke is on surer ground. In 1968, he anticipated how computers would communicate with their masters - not by printer but by visual display years, one ape among many acquired reason, language, self-consciousness and all the properties of mankind. We gained the opposed thumb of the toolmaker, the huge brain and difficult birth that go with a big-headed baby, the vast time-extension that is human childhood, the single human family as opposed to the succession of rapidly maturing young of the typical mammal, and the mental capacity and even need for language. We also gained the snags and bugs that go with such rapid development, unlike the polished perfection of more limited but more slowly evolving creatures. Aliens did it all, says Clarke. The film has a wonderful shot in which an apeman throws a bone, the first murder weapon, into the air. It is immediately followed by a spacecraft – another object thrown by mankind into the air. Indeed, the film's mysterious ending seems to suggest that those unseen aliens are preparing humanity for its next step.

2001 has not confirmed Clarke's biology. Human beings cannot yet be frozen into hibernation, and later warmed back to life. Some bold pioneers hope to make it work, especially those with now-lethal diseases that may be curable in the future. At present, the initial uniform cooling is more troublesome than the later reheating. Similarly, centrifugal force makes a poor substitute for true gravity. The wheel-like space station and the internal rotor on Discovery have not come true. Even the record-breaking Russian

2001 has not confirmed Clarke's biology. Human beings cannot yet be frozen into hibernation, and later warmed back to life

stays in space have not exploited centrifugal force. And the minor biological jokes of the film, the zero-gravity toilet and the special space rations, still give us a lot of trouble.

The economics is even worse. Space technology in the real 2001 would have disappointed Kubrick. We have not built his big spinning space station; even its precursor is too expensive for us. The shuttle to the moon and its Clavius base are still far in the future, if they ever get made at all. The sheer cost of carting their raw materials, or extracting them locally, is daunting. Discovery itself would be absurdly costly nowadays. The real world has gone a more humble way. Russian-American rivalry has collapsed, and the Bell monopoly on telephones is over. And no circuit board has yet displayed the verbal fluency and mental flexibility of HAL. That may be just as well.

The mammoth extinction

Tim Radford

Wiping out the woolly mammoth seems like a mammoth task in itself. But it happened. The question puzzling palaeontologists and climate scientists is: who or what did it? Until the late ice ages, 11,000 to 50,000 years ago, vast creatures stalked the continents of North America and Australia, not then settled by *Homo sapiens*. Many of the bulky beasts vanished from the fossil record, at seemingly the same times. Some 35 genera – the cats, from tigers to tabbies, are all one genus – of large North American mammals, including horses, camels, bears, sabre-toothed tigers, mammoths, mastodons and giant sloths, vanished.

John Alroy, an evolutionary biologist at the University of California, Santa Barbara, made a computer model of North American ecology during the Pleistocene era. This model simulated the population dynamics of 41 large herbivores – and the arrival of bands of human hunters starting 14,000 years ago. Mass extinctions, he reported in *Science* in June 2001, were unavoidable. Once new predators arrived, the biggest creatures, which had the slowest growth rates and the longest gestation periods, were most at risk as populations began to fall. In historic fact, as opposed to computer simulation, about two thirds of all the large American mammals perished around this time.

Dr Alroy said: 'I think we have this problem nailed.' He got implicit support in the same issue of *Science* from Tim Flannery of the South Australian Museum in Adelaide, and Richard Roberts, a geochronologist at the University of Melbourne. Flannery and Roberts used sophisticated dating techniques to link rock ages with animal remains, and dated the graves of large animals at between 51,200 and 39,800 years ago – just as human beings were spreading across Australia. Altogether 28 genera and 55 species of vertebrate are thought to have vanished from the Australian landscape at around this time, and these include giant claw-footed kangaroos and Genyornis, the biggest bird ever to stalk the land. Humans would have brought with them fire, which delivered an indirect blow by changing the vegetation, as well as the usual blows with clubs and throwing sticks. 'If humans had not arrived in Australia, the megafauna would not now be extinct,' said Dr Roberts.

Not everybody is keen on the 'overkill' theory of ancient extinction. Some prefer to think humans might have introduced a new virus or 'hyperdisease' to the American continent. Some point to evidence at Monte Verde in Chile that humans may have arrived in the Americas 33,000 years ago. Some wonder why the great American extinctions actually stopped about 10,000 years ago, if humans were to blame. In the *Journal of World Prehistory*, Donald Grayson of the University of Washington proposed climate change as the real culprit. He said that the theory that hunters 11,000 years ago caused the extinctions was 'glitzy, easy to understand and fits with our modern image of ourselves as all-powerful. It also fits well with the modern Green movement and the Judaeo-Christian view of our place in the world.' But, he said, 'it has now become something more akin to a faith-based policy statement than to a scientific statement about the past.'

If humans had not arrived in Australia, the megafauna would not now be extinct

Damien Hirst's infamous installation of an embalmed calf preserved in formaldehyde.

The story in the stones

Henry Gee

It was the year of whales with dainty ankles, dinosaurs with feathers, and hominids with human faces

Perhaps appropriately, 2001 has been a good year for Missing Links, when, proudly dressed as commemorative gorillas, we throw the thighbone heavenwards, à la Stanley Kubrick, with such force that it becomes a spaceship. There have been whales – 45-million-year-old fossils of whales with spindly legs and hooves that would have been as ocean-going as beachcombing Alsatians (to judge from their diets, they appear to have been salty sea dogs indeed). There have been yet more fossils a bit further seawards, looking less like sei whales than sea lions, equipped with sturdy, flipper-fringed legs of a kind you never see in whales nowadays. Surely, a pair of missing links between the land and the sea.

Then there have been discoveries closer to our own hearts: unearthed from Kenya, a fossilized human that lived 3.5 million years ago, with a brain appropriately primitive for the age, welded on to a startlingly modern face. If ever there were a Missing Link, this would be it, the *sine qua non*, the *primus inter pares*, the *crème de la crème*. Except, that is, for one awkward fact.

The problem is that, outside the minds of advertising copywriters and harassed news editors, 'missing links' are phoney baloney. A moment's thought will show why the term 'missing link' should be condemned to the junk heap of outmoded concepts, along with phlogiston, the luminiferous aether and Lieutenant Uhura's hemline. At root, it is a metaphor for evolutionary history. At one level, we can think of a link simply as a segment of a chain – a chain of ancestry and descent. But when headline writers describe missing links, they mean some hitherto undiscovered, extinct species that is intermediate in form between two others that are already known. The most popular usage refers to human evolution, to describe fossils that appear human in some respects, but ape-like in others. The implication is clear – human beings evolved from apes, and the newfound fossil provides a link between the two.

Most people would find this analysis uncontroversial. To expose the depth of its error, I shall need to pursue the metaphor a little further.

All modern biology is based on Darwin's theory of evolution by natural selection. The core of this theory is very simple to understand. Darwin started with the Malthusian assumption that living things tend to reproduce profligately, outstripping available resources. The offspring that manage to reach reproductive age themselves will, therefore, be those most suited – or 'fitted' to the prevailing environment. Because reproduction is accompanied by genetic variation, it is easy to see that the environment will favour some variations over others, with the result that, over generations, the individuals of a species will become better adapted to their environment. This action of the environment on superabundant, varied creatures is called 'natural selection'.

It is easy to see that natural selection is a mindless consequence of disparate forces, with no discrete memory, personality or purpose. If natural selection is depicted as Death with his scythe and cowl, such personification is neither more nor less poetic than Keats' picture of Autumn sitting carelessly on a granary floor. Crucially, the cumulative effects of natural selection over geological timescales – one might say the 'direction' of selection – cannot be predicted on the basis of its day-to-day activity, even in principle. If we can say anything about the results of such effects, it will be something very general: as populations of individuals change and adapt, each to their own circumstances, we should be able to see a bush-like pattern emerging. Darwin's own example is telling – he explained the diversity of the many species of finches on the Galapagos Islands as the effects of natural selection acting, over millions of years, on a small population of finches that arrived on the islands from South America. Each modern species had evolved separately, according to its needs, from the same common ancestor.

And yet, in the popular mind, Darwinian evolution has become inelegantly welded on to pre-evolutionary ideas of the so-called *scala naturae* – the ladder of nature – in which each organism had its station in a parade of creatures arranged in a linear fashion according to their relative closeness to the human form. Scripture dictated that the human form was the closest to the divine: thus, the *scala naturae* was given direction and purpose. Human beings form a rung – a link – between the apes and the angels. To many, natural selection draws arrows between these separate stations, allowing evolutionary progress and – importantly – improvement. This notion of improvement is deeply ingrained in the public mind as a central feature of evolution. We see humans as more advanced than apes in the same way as advertising copywriters tell us that the latest model of car is superior to last year's version. 'It's evolved' ran the payoff line in a recent car

And then came the key discovery, in Pakistan, of the first whale with legs – Ambulocetus, the 'walking whale'

commercial, the voice-over provided – ironically – by a prominent geneticist. Yet the idea of improvement is profoundly antithetical to the Darwinian conception of evolution.

From this, you can start to see why the term 'missing link' is, at best, inappropriate, and at worst, a kind of bastard voodoo cult in which the new religion is superimposed on the old gods. But there's more. So far, I have discussed the bastard origins of the word 'link', but this is just one half of the phrase at issue. To know that a link is 'missing' we have to be aware of a pre-existing gap in the chain that is waiting to be filled. Again, such foreknowledge is an affront to the fundamental property of natural selection – that it is directionless and free to act on circumstances as they change from moment to moment. If it is anything, natural selection is not simply a motor that drives evolution along pre-set rails.

From this, it should be clear that the term 'missing link' has no place in rational discourse. The public deserves better. Let's look at those whales again. Modern whales are so perfectly adapted to ocean life that it is hard to imagine them as having evolved from creatures as fond of terra firma as you or I. Most ancient Greeks and all small children think whales are fish. Whales have gone to great pains to disguise their past. Whereas all land mammals have a leg at each corner, each fringed with fingers or toes as appropriate, whales have a pair of flippers at the front, and the hind limbs have atrophied to minuscule splints of bone buried in the body wall. But whales retain distinctive vestiges of land life. They are mammals – giving birth to, and suckling, live young – and all the evidence suggests that mammals evolved on land. Somewhere, sometime, whales evolved from land mammals with a penchant for paddling. But which ones? For want of anything else, palaeontologists pointed their fingers at the so-called 'condylarths', a bunch of primitive creatures so motley that they could have evolved into almost anything. Sometimes dubbed 'archaic ungulates' – a term that should bring to mind, if anything at all, the racehorses in the paintings of George Stubbs – some of these shambling brutes were distinctly carnivorous. One of them, Andrewsarchus, had a metre-long head full of huge teeth and looked like the Hound of the Baskervilles on steroids. Hardly the peaceful grazer.

But light dawned with the idea that whales could have shared a common ancestry with some thoroughly modern mammals – the artiodactyls or even-toed ungulates that today include cows, sheep, pigs, deer, camels and hippos (but not horses). Artiodactyls have shapely and utterly distinctive ankles. Fossils of a giant extinct whale called Basilosaurus – a 50-foot monster from Egypt that was the closest thing to a sea serpent that the Earth has ever seen – were found to have tiny, vestigial hind limbs that looked very cow-like. And then came the key discovery, in

It is a failure of logic to assert that one particular fossil was the ancestor of the next one in the chain. The reason is easy – no fossil is ever buried with its birth certificate

Pakistan, of the first whale with legs – Ambulocetus, the 'walking whale', equipped with strong hind as well as forelimbs.

Although evidently adapted for an aquatic life – it was rather like a giant otter – its existence a far cry from the completely seagoing existence of modern whales. Ambulocetus was a Missing Link. In 2001, Ambulocetus is joined by Rodhocetus – another whale in transition, this time resembling a sea lion, though with very cow-like anklebones and distinct hooves on its flippers. To complete the picture, researchers have found the skeletons of Pakicetus and Ichthyolestes, two primitive whales previously known only from skulls. These two dog-sized creatures had cow-like limbs terminating in hooves.

It would be tempting to arrange these fossils in a neat line, like links in a chain, from land-living Pakicetus and Ichthyolestes, to otter-like Ambulocetus, sealion-like Rodhocetus, ocean-going Basilosaurus, and finally to a whale capable of inspiring someone to say 'Call me Ishmael', as in Melville's *Moby-Dick*. This is a temptation we must resist. (The arrangement of fossils in lines, that is.) Of course, whales must have evolved from land animals, and this sequence of fossils gives us a clue to what the transitional stages would have been like. But it is a failure of logic to assert, as if it were certain knowledge, that one particular fossil was the ancestor of the next one in the chain. The reason is easy to see – no fossil is ever buried with its birth certificate.

Just try this simple thought experiment. Let us say you went to the African desert in search of one of your fossilized ancestors. This seems entirely reasonable. After all, you had ancestors, and there is good evidence that humanity began its evolutionary journey in Africa, so it makes sense that some creature, preserved as a fossil, is (or was) your lineal ancestor.

If you came across a likely-looking fossil, though, how would you know whether or not it was your ancestor? It is not as if a fossil springs enlivened from the ground, like something out of the Book of Ezekiel, shrieking 'Ishmael! Where have you been all these years? Come and have a bagel!' Fossils (perhaps wisely) are mute. It is up to us to tell their stories – stories full of ancestry and descent. In the same way, there is no way to tell whether one fossil whale in your sequence is the ancestor of another or not. The most you can say is that they are all cousins, each a twig on the tree of life envisaged by Darwin; each a creature in its own right, complete and adapted to its surroundings, whose reason for existence is simply to reproduce, and not to be some staging post in some vast preordained journey.

The evolution of birds from dinosaurs is another, even older story. From a palaeontological viewpoint, it began just two years after Darwin published the *Origin of Species* and has been rumbling away ever since. Over the last 20 years, the story has picked up pace with renewed vigour, coming to a head in the late 1990s when it was realized that birds make up just one of several groups of dinosaurs with feathers.

Palaeontologically speaking, birds are special because they are the only dinosaurs alive today, but it is possible that the birdlike habit evolved several times. As the palaeontologist and dinosaur artist Greg Paul has noted, it may be that some dinosaurs thought to be completely terrestrial in their habits were descended from flying ancestors. Was Tyrannosaurus rex a flying dragon that fell to Earth? At the moment, nobody suggests any such thing, but the possibility is there. Significantly, this imaginary possibility doesn't exist in any world-view that sees evolution as progressive, directional and predictable – in other words, an evolution made of missing links.

Contemporary critics of Darwin's *Origin of Species* pointed out that were evolution true, we should see evidence of transitional forms, creatures somehow intermediate between extant groups of animal. We can now see that this expectation is based, to a degree, on the false premises that underlie the concept of the missing link. Nevertheless, a suitable candidate turned up in 1861, just two years after the *Origin* was published. This was Archaeopteryx ('ancient wing'), the 150-million-year-old fossil of a bird fully clad in feathers, but with teeth instead of a beak, and many other reptilian features. Archaeopteryx was the 'missing link' between birds and reptiles, in particular the then-recently discovered group of extinct reptiles called 'dinosaurs'. In the late 19th century, birds and dinosaurs were seen as close relatives, but this tendency ground to a halt thanks to work in the early years of the 20th century showing that dinosaurs lacked the fused clavicles or 'wishbones' that are a prominent feature of birds. After that, the many similarities between birds and dinosaurs were seen as parallelisms, and the search began for an ancestry of birds that explicitly accounted for the origin of flight, and, as an important corollary, the origin of feathers.

Unfortunately, the requirement that the origin of birds must also account for feathers rests on rank illogicality. True, all birds (including Archaeopteryx) have feathers. But it does not follow from this statement that all animals with feathers must be birds. This is as ridiculous as saying that because all giraffes have four legs, then all animals with four legs must be giraffes.

Beginning in the 1970s, some researchers began to wonder whether the close similarities of birds and dinosaurs were more than just coincidence. True, dinosaurs didn't have feathers, but they had many other birdlike features, such as fused and hollowed-out limb bones. Some dinosaurs were shown to have had wishbones – overturning half-century-old received wisdom.

And then came dinosaurs with feathers. In the 1990s, the Western world learned of a series of

Despite the shock of its human face, this creature – Kenyanthropus – was no more human than any other creature of its time

remarkably productive quarries in Liaoning province in northeastern China that yielded immense numbers of incredibly well preserved fossils. In 150 years, the Bavarian limestones that produced Archaeopteryx have given up only seven specimens. In less than ten years, the rocks of Liaoning have produced hundreds, if not thousands, of specimens of Confuciusornis, a primitive bird complete with feathers and horny beak. The Liaoning beds have also produced tiny mammals still in their furry coats, and many other spectacular specimens. To date, it has produced specimens of six different types of dinosaur, all of which have been clothed in a kind of fibrous coat that seems to be the earliest evolutionary manifestation of feathers (imagine T. rex rolling around in the shag pile and you'll get the idea). Two of these Chinese dragons, Protarchaeopteryx and Caudipteryx, also had proper feathers, recognizable as such – the former as a small switch at the end of the tail, the latter with a tail-plume as well as go-faster fringes on its arms. These creatures almost certainly could not have flown, and were not especially close relatives of birds. The age of the Liaoning fossils is contentious, but estimates range from around 124 to 145 million years. If the very earliest Liaoning fossils were younger than Archaeopteryx, some critics have wondered, how can some of the feathered dinosaurs of Liaoning have been precursors – or ancestors – of birds? This question is based, once again, on the false premises of the missing link. What we are seeing is not a linear series of fossils leading from dinosaurs to birds, arranged according to their flight capability, but the evolution of feather-like structures in a range of dinosaurs, some of which used their feathers to fly, and others that did not. The implication is that some animals we think of as dinosaurs could have had flying ancestors. Now, isn't that a more mind-expanding alternative?

Of course, the term 'missing link' is classically applied to stories about the evolution of our own species. It started in 1925 with the discovery of the skull of an 'apeman' in a quarry called Taungs in South Africa. The discoverer was one Raymond Dart, who called his creature *Australopithecus africanus* – the southern ape from Africa. Given that Darwin had speculated on Africa as the cradle of humanity, *Australopithecus* was seen as a transitional form, the archetypal missing link between ancient and modern. Africa remained a magnet for the curious interested in unearthing further missing links in a predictable evolutionary chain between ape and man. The problem was that Africa was not so obliging: each new discovery raised hitherto unsuspected questions, and adamantly refused to be slotted into a pre-existing gap.

By the 1990s, it was clear that human evolution was more like a bush than an easily tractable linear series. A picture was emerging in which hominids – members of the human 'family', of which our own species *Homo sapiens* is the only survivor – appeared in Africa around 6 million years ago. By 3 million years ago, there were three – possibly more – different hominid forms living in Africa. One was *Australopithecus africanus*. Another was a group of specialist plant-eaters called *Paranthropus*. The third were the earliest members of our own genus, *Homo*.

The search for human ancestry has long been associated with the illustrious name of Leakey, and it was in the 1970s, on the eastern shore of Lake Turkana in Kenya, that one Bernard Ngeneo, one of Richard Leakey's associates, discovered a skull known only by its catalogue number, KNM-ER 1470, or '1470 Man'. The handle is prosaic, but the skull itself can never be, for it is the earliest known example of a face that is recognizably human. To look at the face of 1470 Man is to be reminded of the archaeologist Schliemann confronting the death masks of Mycenae and wondering whether he had gazed on the face of Agamemnon. 1470 Man is approximately 1.8 million years old and is usually ascribed to a very ancient member of our own lineage called *Homo rudolfensis*.

To unearth a fossil and find, staring back at you, something as distinctive as the human face, is to discover ancestry, something that surely must qualify as a missing link. But not even such cherished marks of humanity are immune from logic, nor from the fecundity of nature, which will always outsmart us when we think we have drawn the major features of evolution so well that we can define gaps in which to slot our missing links. In 2001, Meave and Louise Leakey (Richard's wife and daughter) and their colleagues described a creature from the western shore of Lake Turkana that was around 3.5 million years old. Although startlingly primitive in many ways, as one would expect for a creature of such antiquity, it had a face disconcertingly like that of 1470 Man, at least at first glance. This new creature cannot easily be fitted even into the prevailing tripartite model of human evolution and seems to represent an entirely new form. Despite the shock of its human face, this creature – Kenyanthropus – was no more human than any other creature of its time.

A lazy news editor could easily hail Kenyanthropus as a missing link, but now we know better than to fall into that trap. Kenyanthropus represented just one of an unknowable number of different ways of being human. It was its own creature, doing its own thing, not a stepping stone between apes and angels – and a graphic demonstration that the way of thinking represented by the term 'missing link' is as outmoded in today's world as the one in which Earth was flat, static, and only 6,000 years old. We live in a universe far vaster and stranger than that dreamed of in that ancient philosophy, and that strangeness and wonder applies to evolution, too.

The tsunami hazard

Tim Radford

One of the worst disasters of all time has yet to happen. It will be a giant Atlantic wave that could smash into British coasts, damaging power stations and flooding low-lying regions. But the British would be among the lucky ones. Waves the height of Nelson's Column in London would probably hit coasts from Canada to Brazil, sweeping up to 10 miles inland and causing unspeakable destruction.

Simon Day of the Benfield Greig hazard research centre at University College London and Stephen Ward of the University of California worked out what it would take to trigger the megatsunami. They reported in *Geophysical Research Letters* that one day an eruption of the Cumbre Vieja volcano in the Canary Islands would send a lump of rock twice the volume of the Isle of Man sliding at 200mph down the unstable western flank of the mountain. Hitting the sea with an almighty splash, the rock would tumble for at least 40 miles along the seabed. This would set up a tsunami or wave train involving the entire water column, rather than the top few metres usually whipped up by winds. Tsunamis characteristically gather speed according to ocean depth. They can cross the deepest oceans at 500mph, slowing but building up into terrifying bodies of water when they reach coastal shelves. 'The first wave is going to come in, maybe take out the first few blocks, take the debris away, flatten the ground. The next wave takes out blocks progressively further inland. Over a large part of the area that is inundated, you will be seeing near-total destruction,' Dr Day said.

Tsunamis are thought to have killed about 4,000 people during the last decade. In the past 100 years, around 50,000 people have been swept away by 400 tsunamis generated by earthquakes on the Pacific rim alone. Volcanic collapse is a relatively rare trigger for tsunami destruction, but geophysicists have now identified 11 cases of

> Waves the height of Nelson's Column would hit coasts from Canada to Brazil, sweeping up to ten miles inland and causing unspeakable destruction

massive landslides to the ocean floor in the past 200,000 years, in the Hawaiian islands, Cape Verde and the Canary Islands. After an eruption in 1949, an ominous fissure appeared in Cumbre Vieja, first evidence of a huge fault line in the massive basalt rock of the mountain, which could one day turn into another landslide.

In the same month as Day and Ward's article, there was separate confirmation of the hazards posed by submarine landslide anywhere in the world. In August, engineers told a tsunami conference in Istanbul that, according to computer models, such an event must have engendered the tsunami that smashed into Papua New Guinea in July 1998, killing more than 2,100 people. The 60-foot waves smashed every structure on a 15-mile sand spit and impaled bodies on tree branches, probably because of a slump of rock and sediments about 15 miles from the stricken coast.

Further back in time, the collapse of a volcano on the Greek island of Thera (now Santorini) in about 1600 BC set up waves that would certainly have shaken Minoan civilization, though not forced it to collapse. As Costas Synolakis of the University of Southern California in Los Angeles told *Science* (17 August), these waves might have been 360 feet high to begin with: 'The size of the wave that actually reached Crete would have been disruptive, but it would not have devastated the Minoans to the point that they abandoned their palaces.' In other words, the Minoans were lucky, that time. Many of the 60,000 who died in the Lisbon earthquake of 1775 actually perished not in the initial shock, but in the resulting tsunami. But for most of human history, tsunamis have claimed lives in the Pacific rather than other oceans. In 2001, disaster scientists showed that, ultimately, almost any coastal city, anywhere, could be at risk.

Cancer: it's all in the genes

Mike Stratton

Cancer is simply DNA gone out of control

Approximately one in three people in Europe and North America develops cancer. One in five will die of the disease. All cancers arise due to abnormalities in the DNA sequence of key genes. If we knew which these genes were, we could begin to understand how cancer develops, think of new ways to treat it, and new ways to prevent it. The Cancer Genome Project aims to use the human genome sequence to search systematically through the genomes of cancer cells to find the abnormalities and identify the critical genes implicated in human cancers.

In the adult human body there are about 100 million million cells. During development from the fertilized egg to the adult, each cell specializes in performing particular functions. Some become muscle cells, nerve cells, fat cells, immune cells and so on. Despite this diversification of function, each of the 100 million million cells contains a copy of the 6,000 million base pairs of DNA (3,000 million contributed by the sperm of the father and 3,000 million contributed by the mother's oocyte) that were originally brought together in the fertilized egg from which the individual developed.

During the lifetime of the human being, the DNA of each cell is continuously bombarded by chemicals, irradiation and viruses. In addition, the remarkably reliable machinery that copes with the enormous task of faithfully replicating each of the 6,000 million letters of DNA code during many thousands of cell divisions occasionally makes mistakes. As a result, the DNA sequence of each cell in the adult differs very subtly from that of every other cell in that person. The genome in each cell is still more than 99.99 per cent identical to the genome in any other cell from that person and to the original fertilized egg. Nevertheless, these occasional differences (known as somatic mutations) do accumulate throughout life. Most do no harm.

Occasionally, however, a single cell acquires a

A central aim of cancer research has been to identify these critical mutated genes that provide the 'hard wiring' for the abnormal behaviour of cancer

somatic mutation in an important gene (perhaps one that normally controls cell division), which now causes that cell to divide a little more frequently than it should. After several rounds of cell division, the cells that derive from this single mutated cell (known as a clone of cells) may now be a little more numerous than they should be, and take over a small area of the tissue in which they are located. This sort of thing is probably happening all the time in the human body, with many hundreds of these small, abnormal microscopic clones of cells more or less peacefully coexisting with normal tissues and still causing no obvious trouble to the health of the person.

Sometimes a single cell within one of these many expanded clones may acquire another mutation in a different critical control gene. Perhaps this time the mutated gene controls cell survival. As a result, the programme of normal cell death that the cell usually adheres to rigorously is now subverted, and cells of this clone survive longer than they should. This new clone of cells with two critical mutations expands a little more within the tissue and perhaps begins to distort and disturb the microscopical architecture around it. Several rounds of this process of random somatic mutation (we do not know precisely how many), followed by expansion of selected clones of cells, are required before a single cell acquires the ability to ignore completely the normal organizational rules of the tissue around it, to invade surrounding tissues and ultimately to be carried off by the bloodstream and deposited in another organ. This cell and its progeny are a cancer.

A central aim of cancer research has been to identify these critical mutated genes that provide the 'hard wiring' for the abnormal behaviour of cancer. Indeed, following the invention of technologies that allowed manipulation of pieces of DNA (known as recombinant DNA technology) more than 20 years ago, there has been much success in the identification of cancer genes. These genes and their protein products control a variety of cellular functions including repair of DNA, cell-to-cell contact and cellular differentiation, in addition to cell division and cell death. However, there is strong evidence that we have not found most of them. In particular, for common adult tumours such as breast or prostate cancer, we have scarcely scratched the surface.

The advent of the human genome sequence offers enormous potential for developing new approaches to tease out the cancer genes. Previously, finding these genes was a very laborious process. First, we had to work out (by a variety of strategies) approximately which small region of the genome contained a critical cancer gene (such regions might contain tens or even hundreds of genes). Second, we had to provide the DNA sequence of this region ourselves. Third, we had to identify which of the genes within the region was mutated in cancer and was therefore the one we were looking for.

Now that the sequence of the whole genome is available to us, the second step in this process is no longer needed. Even more important, however, is that instead of worrying about the first step (i.e. locating the approximate position of the cancer gene in the genome), we will be able to search systematically, gene by gene, through the genomes of cancer cells, comparing the DNA sequence to that in normal cells. Where we find differences in DNA sequence between the cancer cell and the normal cell, these will highlight the presence of a cancer gene. This is the primary aim and programme of the Cancer Genome Project.

Although almost ridiculously simple in conception, it is a little more taxing in implementation. There are many reasons for this, but by far the most important is the sheer size of the genome. With 30,000–35,000 genes, we estimate that we will need to do 400,000 experiments using our current technologies just to look at all the genes in a single tumour sample. Given that there are approximately 100 distinct classes of cancer and that even within cancers of the same type there is considerable diversity with respect to the genes that are mutated, there are tens of millions of experiments to do even to obtain a superficial perspective on human cancer.

We should not be too daunted by these numbers. Advances in technology may render these challenges facile. In many ways, we are in a similar position now to that of genome sequencers 10–15 years ago. Then, there was intense debate over whether to wait for a new process that would allow relatively quick and cost-effective sequencing of the enormous human genome, or whether to make a start using the conventional sequencing approach invented by Fred Sanger and to scale up gradually to a position from which the genome could be tackled as a whole. The latter approach ultimately delivered the result.

In the Cancer Genome Project we have similarly decided not to wait. We are souping up the currently available technologies to a level at which cancer genomes can be compared to each other and to normal genomes. That is the reason why the Cancer Genome Project has been sited at the Wellcome Trust Sanger Institute, the British genome centre at Hinxton, near Cambridge. Genome centres have the modern robotics, the infrastructure and the massive computing power to carry out millions of experiments at high speed, track the results and analyse the output – exactly what we need to implement this ambitious plan.

First, however, the draft human genome sequence needs to be finished – that is, there should be fewer than one in ten thousand erroneous bases and as few gaps as possible. This will be achieved within 18 months to two years. Then the positions of all the genes will need to be identified (a process known as annotation), so

It would be very surprising if, in 20 years' time, there had not been a very substantial improvement in cancer treatment

that we can focus our analyses on them, rather than wasting time on the junk DNA between genes. This will probably happen by the end of 2004. Overall, we expect the Cancer Genome Project will take a decade to provide information on the full spectrum of human cancer.

The identification of the mutated cancer genes is not the end in itself. Such discoveries lead to novel lines of research that provide new insights into the mechanisms by which cancers develop. The somatic mutations can even serve as archaeological records of the causes of a cancer, the environmental factors that were assaulting the genome of each cell in the body decades before the cancer itself emerged. However, by far the most important reason for finding these genes is that they are the targets for modern drug development. Since these mutated genes are the hard wiring of the cancer cell, switching off their effects might mean quelling the mutinous behaviour of the cancer cell itself.

Is the notion of drugs targeted at key mutated genes an elegant but wishful fantasy? Until recently, perhaps, it was an article of faith amongst cancer scientists. However, we now have at least one example in which it has yielded remarkable results. In an adult cancer known as chronic myeloid leukaemia (CML), a drug has been developed against a gene that is mutated in almost all cases of this disease. In the first stage of trialling, 53 out of 54 patients who had failed to respond to all conventional treatments for CML went into long-term remission, with very minor side effects. Not surprisingly, the drug has quickly been licensed for routine use.

Such extraordinary advances are rare in cancer research and the community of scientists is buzzing with the news. We should, however, retain a sense of perspective. Not every mutated gene will lend itself to being targeted by drugs. Not every drug against a mutated gene will be as effective. Some drugs that work well may be associated with serious side effects. Most importantly, cancers are cunning and remorseless adversaries that retain many hidden weapons in their armoury, allowing them to sneak away and put up determined resistance. Nevertheless, it would be very surprising, if in 20 years' time, there had not been a very substantial improvement in cancer treatment. Much of this will be based on the use of the human genome sequence in cancer research.

The dinosaur's nose

Tim Radford

Old bones continued to make news stories. Some finds were felicitous. Josh Smith of the University of Pennsylvania set himself the PhD challenge of tracking down some Egyptian palaeontological sites recorded by the German scientist Ernst Stromer von Reichenbach, whose Munich collection was destroyed by bombing in 1944. Smith told *Science* on 1 June that when he set off for the Bahariya oasis, where Stromer had apparently made rich finds, he put the wrong co-ordinates in his global positioning system receiver. So he ended up some way from the expected landmark. To get his bearings, he looked out of the passenger window of his Toyota Land Cruiser and immediately spotted a large dinosaur bone. In three weeks, Smith and colleagues excavated 6.5 tons of a titanosaurid that might have weighed 100 tons, and which must have waded through Cretaceous swamps with feet a metre across. He called it *Paralatitan stromeri*.

In Pakistan, sifting bones from a bank of ancient sediment, Hans Thewissen of Northeastern Ohio University's College of Medicine assembled the outline skeleton of *Pakicetus attocki*. This was the size of a wolf, with the anklebones of a bovid but the earbones of a whale. This splashing predator was reported in *Nature* on 20 September as fresh evidence of a lineage that evolved on land but was already heading back towards life in the sea, and a future as a deep-sounding, krill-straining, blubber-insulated cetacean. In the desert of Niger, Paul Sereno of the University of Chicago – with a number of remarkable finds already to his name – unearthed the skull of a Cretaceous crocodile 40 feet long and weighing eight tons. Its diet, he reported in *Science* on 26 October, must have included not just fish but small dinosaurs. Sereno and his colleagues were puzzled by the sheer size of the creature's nose.

Meanwhile, at Ohio University's College of Osteopathic Medicine, Lawrence Witmer completed research that could literally change the face of dinosaur studies. After X-raying and dissecting the soft tissue nasal structures of modern crocodiles, lizards and birds, he took a new look at the standard artists' impressions and movie representations of the great beasts of the Jurassic and Cretaceous periods. Their noses were, he decided, placed far too high on the head and back from the snout. Properly, the fleshy nostrils should be further forward and closer to the mouth. Noses were clearly important to dinosaurs. Half the volume of the skull of triceratops was given over to nasal cavity. 'The nostril project was one I was almost scared to get into,' he told *Science* on 3 August.

The antimatter conundrum

Frank Close

A flicker from a fossil relic in 2001 could provide the answer

Antimatter has an aura of mystery, the promise of a natural Tweedledum to our Tweedledee, where left is right, north is south and time runs in reverse. Its most celebrated property is its ability to destroy matter in a flash of light, converting the stuff that we are made of into pure energy. In science fiction, antiplanets tempt travellers to their doom even as antihydrogen powers the engines of astrocruisers. In science fact, according to everything that decades of experimental physics have taught us, the newborn universe was a cauldron of energy where matter and antimatter emerged in perfect balance. That is, light, or energy, condenses into equal quantities of matter and antimatter, with an antimatter match for every form of matter. Which begs a question: how is it that matter and antimatter did not immediately destroy each other in an orgy of mutual annihilation? How is it that today, 15 thousand million years later, there is anything left in the universe at all?

This conundrum touches on our very existence. We are made of matter, as is everything we know of in the universe. There are no antimatter mines on Earth, which is just as well, as they would be destroyed by the matter surrounding them, with catastrophic results. Somehow, within moments of the Big Bang, matter had managed to emerge victorious: the antimatter having been annihilated, the heat energy from the destruction remained (now a cool 3°C above absolute zero and known as the microwave background radiation), and the surfeit of matter eventually clumped into galaxies of stars, suns and everything else, including the *Sun* and the *Guardian*.

What aspect of nature caused this lopsidedness, this leaning towards matter to occur? This question has plagued physicists and cosmologists for years. But in 2001, an important clue was found, which was the first major advance in this particular area for some 30 years. Results from two experiments, one in Stanford, California and the other in Japan, and with physicists from several British institutions among the participants, may

How is it that matter and antimatter did not immediately destroy each other in an orgy of mutual annihilation? How is it that today, 15 thousand million years later, there is anything left in the universe at all?

> When the team was awarded the Nobel prize for physics, a Swedish newspaper announced that it was for 'the discovery that Nature's laws are wrong'

eventually prove to be a breakthrough in this quest. The saga is not quite like seeking a needle in a haystack, because an essential clue turned up 37 years ago and it is only now, following further discoveries and advances in technology, that it has become possible to exploit the clue. The clue was the discovery that nature contains a tiny imbalance, a tendency for the behaviour of certain 'strange' particles, known as K-mesons, not to be precisely mimicked by their antimatter counterparts.

The strange particles do not constitute the stable matter that we are used to on Earth, but are produced in violent processes in the cosmos. They were discovered in 1947 among the debris arising when cosmic rays hit the upper atmosphere. The realization that there is exotic stuff in the universe helped to inspire the building of particle accelerators, which were capable of producing strange particles, such as K-mesons, in abundance. These ghostly particles have very short lives, before they decay. Thus it was that in 1964 a team of physicists in New York discovered that about one in every ten million times, the matter and antimatter accounts in the K-meson decays failed to balance. When the team was subsequently awarded the Nobel prize for physics, a Swedish newspaper announced that it was for 'the discovery that Nature's laws are wrong'!

This asymmetry is so small that investigating it has been one of the most demanding and delicate measurements in modern physics. The results from decades of study, with ever-increasing precision, suggested that there might be another place in nature where this type of asymmetry exists, and much more dramatically than in the only one known so far. In 1977, the first examples of 'bottom' particles were discovered at the Fermilab accelerator near Chicago. It turned out that the previously unknown bottom particles are in effect heavier versions of strange particles. As the strange particles appeared able to distinguish between matter and antimatter, so might the bottom particles. Indeed, when the existence of these bottom particles was incorporated into the existing theory of particles and forces, the resulting equations surprisingly seemed to imply that an asymmetry between matter and antimatter was almost inevitable. Could the bottom particles somehow hold the key to the conundrum? As bottom particles were abundant in the first moments of the universe, might they hold the secret of how the lopsided universe, where matter dominates today, has emerged?

To find the answer, it was necessary to make billions of the ephemeral bottom particles and their antiparticle counterparts, and to study them in detail. To do so, 'B-factories' – customized accelerators that are specially tuned to produce this exotic stuff – were designed and built in California and in Japan. These are relatively compact machines for modern particle physics, being only a few hundred metres in circumference, but involving high-intensity beams of current controlled with greater precision than ever achieved before.

The accelerators were completed in 1999 and, after initial testing, began to collect data. To get definitive results requires creating and studying vast numbers of the bottom particles. It is like tossing a coin: chance might make it come up heads five or even ten times in a row, but if this continues to happen, then something is special about the coin. So it is with the study of ephemeral subatomic particles. They live for less than the blink of an eye and it is what remains after they die – their fossil relics if you like – that have to be decoded. One needs to have huge numbers of such fossils in order to tell whether any differences are real or simply the result of chance.

There are many varieties of fossils that can be studied, and specialist teams at the two accelerators have begun to collect and measure the characteristics of several of these. Among them is a particular species, known as the 'psi-K-short' events, that theorists predicted would be the most immediate indicator of a difference between bottom matter and bottom antimatter. By the end of the year 2000, definite hints of such a difference were being seen in these 'psi-K-shorts', which suggested that the searchers were indeed on the right track, although it was not until the summer of 2001 that the results emerging from the Californian and Japanese experiments finally matched. It is now clear that the first fossil species – the 'psi-K-shorts' – shows a large difference between matter and antimatter, in accordance with what had been predicted. Over the next three years, other fossil types will be examined to see if a common story emerges, or whether more subtle phenomena will be revealed. As to whether this will indeed be the answer to the mystery of matter for the whole universe – that's the big question.

The big heat

Tim Radford

George Bush didn't like the Kyoto agreement, but the planet sent messages about global warming anyway

The Kyoto agreement to begin limiting carbon dioxide emissions into the atmosphere almost collapsed in 2001, and was saved only at the last minute. The intergovernmental panel on climate change began the year by announcing that there had been a 20 per cent decrease in snow cover, and a 40 per cent thinning of the Arctic ice cap, and that without a drastic reduction in carbon dioxide emissions, this warming would continue. Scientists warned last year that temperatures will rise between 1.4°C and 5.8°C in the next century – a more ominous forecast than the one issued in 1996. A rise of nearly 6°C would be higher than any in the past 10,000 years.

In March 2001, the new US president George W. Bush made it clear that the US would not join the Kyoto agreement. He referred to scientific uncertainties and called for advice from his own National Academy of Sciences. He got it. In June the academy issued a report that declared bluntly: 'Greenhouse gases are accumulating in Earth's atmosphere as a result of human activities, causing air and subsurface ocean temperatures to rise.' This put the US researchers squarely in line with the other 17 of the world's most conservative national scientific societies, which in May had called upon politicians everywhere to begin honouring the Kyoto agreement.

There were other, more palpable reminders of a warmer world. In February, British scientists reported on a pattern of thinning of the west Antarctic ice sheet. The continent's biggest glacier, they said, had lost 10 metres of thickness in eight years and retreated 5km inland. In the same month, a US team calculated that the snows of Kilimanjaro and the glaciers in the high Andes of Peru were melting so fast that they could disappear within 20 years. Another US group linked the worldwide deaths of frogs, toads and salamanders to changes in rainfall patterns triggered by global warming. In April, US oceanographers studied 50 years of data from the

Golf spectators in the rain.

oceans to a depth of 3,000 metres, and found that the Atlantic, Pacific and Indian Oceans had warmed on average by 0.06 per cent since 1948. This was more or less precisely what computer models of global warming had predicted would happen as a result of increased levels of carbon dioxide in the atmosphere.

There were other messages from the seas around us. Researchers from Scotland, Norway and the Faeroe Islands measured the flow of dense, cold water across the Faeroe Bank channel and calculated that it had fallen by 20 per cent in five years. This 'ocean conveyor' is an integral part of the Gulf Stream, which keeps Britain at least 5°C warmer than its latitude might dictate. The implication is that the Gulf Stream might one day 'switch off' because of global warming. Paradoxically, a warming world could plunge Britain into a colder future. This would require a steady thinning of the northern polar ice cap, and evidence of this, too, has been consistently confirmed by independent measurement.

The latest studies come from Alaska and Greenland. Since 1917, Alaskans have gambled on the moment of 'ice break': the exact point at which a wooden tripod would fall through a frozen river. This ice break now occurs on average five days earlier than it did in 1917. A Danish and Canadian team studied the data from traverses of the northwest Greenland ice sheet in 1954 and 1995. They found that the ice was thinning faster, and thinning at higher elevations, than anyone had expected.

NASA researchers took a look at the global picture in 2001, matching annual changes in temperature and in vegetation measured by spacecraft over a 21-year period. They found that in Europe and Asia, spring was now arriving a week earlier and autumn was being delayed by 10 days. In Marrakech in November, world environment and energy ministers went ahead without the US – the biggest producer of greenhouse gases – and paved the way for implementation of the Kyoto agreement in 2002. The day before, the US energy department issued its calculation of the amount of carbon dioxide pumped from American car exhausts and chimneys in 2000. It had risen by 3.1 per cent in a year.

Staring at the sun

Tim Radford

Mayan civilization did not collapse so much as wither. It may have desiccated in a cyclic drought that lasted 100 years or more, according to a team sifting through sediments from the bottom of Lake Chichancanab in the Yucatan region of Mexico.

The Mayans farmed maize, beans and pumpkin, were the first to make chocolate, and some people credit them with the invention of the cigar. They built stone reservoirs, invented irrigation systems, erected cities and palaces, and had a mathematical system based on the number 20 (in the ancient world, the Babylonians based theirs on the number 6, while Arabic and then European civilizations later settled on the base 10). The Mayans devised an annual calendar of 365 days, and a ritual calendar of 260 years. Their days, however, were over before the close of the first millennium, long before the arrival of the conquistadors from Spain. And the end may have been because of a harsher than usual drought that seemed to return every 208 or so years.

Researchers reported in *Science* on 18 May that they had examined cores from the Yucatan lake and found a pronounced pattern of gypsum deposits. Whenever rainfall decreased, they reasoned, evaporation from the lake would concentrate salts in the lake water, which would then precipitate calcium sulphate, or gypsum, in a 208-year pattern. This pattern was almost in step with another that astronomers know well: a 206-year cycle of the sun's intensity. Changes in the sun's output worldwide hit the Yucatan region particularly hard, and may have forced a local climate change that burned out a mature civilization. Evidence from the lakebed suggested that the worst drought in 7,000 years set in at about the same time as the Mayan civilization was in decline, and lasted from 750 AD to 850 AD. 'It is ironic that a culture so obsessed with keeping track of celestial movements may have

> But don't blame the sun for everything that is happening now. Variations in solar output have had far less impact on Earth's recent climate than human actions

met their demise because of a 206-year cycle,' said David Hodell, a palaeoecologist from the University of Florida.

Meanwhile, NASA climate modellers confirmed in *Science* on 7 December that a cooler season of sunshine during the Little Ice Age that began almost 600 years ago almost certainly ended the fragile settlement of Greenland. First settled by Norsemen, Greenland was cut off by ice in 1410 and isolated for more than 300 years. The evidence for the intense cold period exists in written records, tree-rings and ice cores. Galileo Galilei began making records of sunspot activity – a crude way of taking the solar temperature – in 1611; others followed, which is why astronomers have a historic record of what is now called the Maunder Minimum. Climate scientists tested a computer model of climate that confirmed an average slight cooling everywhere, with regional variations that would have hit the North Atlantic harder. But don't blame the sun for everything that is happening now. 'Variations in solar output have had far less impact on Earth's recent climate than human actions,' said Drew Shindell of NASA's Goddard Institute for Space Studies. 'The biggest catalyst for climate change today is greenhouse gases.'

Eros on St Valentine's Day

Duncan Steel

One spacecraft landed on an asteroid, and another flew through the coma of a passing comet

Perhaps no group had higher hopes for the year than space enthusiasts: after mulling over the wonderful movie *2001: A Space Odyssey* for three decades, what would the year actually bring? NASA named a Mars probe 'Odyssey' and it joined Global Surveyor in orbit around the Red Planet in October. The space success stories of 2001, though, concerned asteroids and comets. Two American missions delivered the most detailed images ever obtained of the surfaces of such celestial objects.

Near-Earth asteroids (NEAs) are repeatedly in the news, largely because of the hazard they pose to humankind through their occasional catastrophic impacts upon our planet. Connections with the death of the dinosaurs are hard for editors and headline writers to resist. There are many compelling reasons for studying asteroids and comets, but this one is the stuff of nightmares, and Hollywood movies. The first NEA to be discovered, in 1898, was named Eros. Its present orbit does not cross that of the Earth, which rules out any collision in the near future. But it does come close. Computer modelling of its future dynamic evolution under the gravitational tugs of the planets has shown that the paths of Eros and Earth could intersect within the next million years. That is nothing to worry about – there are other NEAs that will run into us before then, anyway – but the huge size of Eros makes this something to contemplate in terms of past mass extinctions of life. This behemoth measures about 21 by 8 by 8 miles, the second-largest known NEA, enough to cause a similar mass extinction if it were to run into our planet. (The only bigger one – Ganymede, found in 1924 – does not come as close, and so is never as bright as Eros.)

Eros was of immediate interest for another reason. By using widely separated observatories and triangulation techniques, astronomers could measure its distance directly, and from that, work out the Earth–Sun separation, a length termed the Astronomical Unit. That then renders the scale of the solar system as a whole, which was a matter of continued argument until radar, a more precise technique for interplanetary distance determinations, was applied from the 1960s. As late as 1975, however, astronomers were still using Eros as a measuring stick. When NASA was planning its Near-Earth Asteroid Rendezvous (NEAR) probe in the early 1990s, Eros was top of the wish list as a potential target. It became the first asteroid to have a spacecraft land on it, in 2001.

Launched in 1996, NEAR was supposed to meet Eros early in 1999 and then drop into an orbit around the asteroid, requiring a rocket burn to decelerate the spacecraft as it approached. A malfunction with the thruster control meant that, instead of slowing down, NEAR went flying on past its intended target. Luckily, the orbits of the two around the Sun were such that NEAR got another chance to meet up again with Eros, a year later in February 2000. This time the slow-down burn was successful.

By that time NEAR had obtained a double-barrelled name. Gene Shoemaker was the doyen of asteroid watchers, and a leader in studies of impact craters. It was he who had taught the Apollo astronauts which rocks to bring back from the lunar surface. In July 1997, whilst on a crater-surveying expedition in central Australia with his wife Carolyn, herself a record-holder for comet discoveries, Gene was killed in a freak head-on automobile accident. Since then his ashes have been delivered to the Moon on board NASA's Lunar Prospector satellite, and a crater in Australia has been given his name. NEAR was renamed; the spacecraft that eventually landed on Eros is called NEAR-Shoemaker.

As befits a tryst with Eros, NEAR-Shoemaker was delivered into a trajectory around the asteroid on Valentine's Day, 14 February. Then it began a convoluted series of loops around Eros that were to continue for just short of a full year. On 12 February 2001, it came to a gentle rest on the asteroid's dusty surface, its mission complete. Along its four-year voyage, the satellite had been directed on an interplanetary trajectory that took it further from us than Eros. In space, the shortest path (a straight line) is rarely feasible. Longer tracks may take more time but require less fuel, and so make better sense financially. The critical factor is the weight of the spacecraft on take-off, and most of this weight is fuel. The more fuel required, the fewer the scientific instruments that can be carried, and therefore long cruise times are often called for. For example the Galileo probe, in orbit around Jupiter since 1995, used fly-bys of Venus and then the Earth (twice) to get the gravitational slingshot boosts that enabled its eventual destination to be reached.

On its way, Galileo made approaches to two

There are many compelling reasons for studying asteroids and comets, but this one is the stuff of nightmares, and Hollywood movies

main-belt asteroids between Mars and Jupiter, Gaspra and Ida. Astronomers classify asteroids according to their colour as seen through a telescope. Those further out from the sun in general tend to be redder than the asteroids closer in, but there are many subtleties in their spectral reflectivities. We get some idea of this from meteorites, which come in many compositional classes: there are several distinct stony varieties and some are purely metallic, while others are mixtures. Asteroid colours seem to fall into similar classifications, as might be expected. Both Gaspra and Ida are stony (or S-type) asteroids.

The rarest meteorites, and the most valuable (both scientifically speaking, and to collectors), are called carbonaceous chondrites. These are crammed full of organic chemicals (molecules based on carbon), hence their name. Their analogues in space are called C-type asteroids. On its long path to an eventual rendezvous with Eros, NEAR-Shoemaker flew past an asteroid in the outer main belt called Mathilde. This is a C-type, and so it could tell us things about the source of the raw materials that made the development of life on Earth possible. As expected from telescopic observations, Mathilde was found to be very black, as dark as soot. A surprise, though, was its low density, much less than that of solid rock. The suspicion is that Mathilde (and probably many other asteroids) is actually a rubble pile containing voids, rendering a low average density.

Eros is quite different, with a reflectivity six times that of Mathilde. Its colour, too, is different: Eros, like Ida and Gaspra, is an S-type asteroid. Overall, the impression is of a large potato-shaped space rock with a hue somewhat like butterscotch. With NEAR-Shoemaker in orbit around Eros for an extended period, researchers were able to obtain detailed images of specific areas they found of interest. While the pictures returned by the camera generate the most excitement amongst the public, other instruments also return invaluable information. For example an infrared radiometer allowed mapping of the thermal properties of the surface, telling researchers whether they were looking at bare rock or a thick, dusty layering, while a gamma-ray spectrometer delivered data pertaining to the elemental composition of the rocks themselves.

The average density of Eros, based on its mass (determined from the force of attraction it exerts on the spacecraft) and its volume (from its size and shape), indicates that, unlike Mathilde, it is a monolith: a single, solid rock. This is confirmed by the large impact craters seen on Eros: some are so large that if Eros were a rubble pile then it would have been demolished by such energetic collisions. Scattered across the surface is other evidence of impacts. Planetary scientists often describe their subjects as having pockmarked faces, pitted by craters, but Eros is also disfigured by a host of warts. Boulders are strewn around, sticking up above the smooth dust-covered profile. Over 30,000 of them have been counted, and a quarter of those are house-sized. While the effect of small impacts upon asteroids is to smooth them, leaving a rounded surface like a pebble, it seems that Eros has suffered at least one major crash. This left a five-mile-wide crater – tentatively named Shoemaker – and ejected rocks that are now spread over much of the asteroid's surface.

Many of the craters on Eros look unusual. They tend to have smooth, flat bottoms, apparently filled by dust that has flowed into them like water. This is called dust ponding. Although it is obvious what has happened, the physical mechanism involved is a matter of debate. The favoured explanation is that dust becomes electrically charged through irradiation by solar ultra-violet light, and then levitates above the surface due to electrostatic repulsion – a phenomenon that has been seen on the Moon – and so can flow like a fluid. The full details have yet to be worked out.

Eros is the near-Earth asteroid about which we know the most. One might say we have seen it all over, warts and all. But during 2001, we also got our best view of the heart of a comet. It had been a long wait. The first up-close images of a comet's nucleus were obtained back in 1986, when five spacecraft flew past the most famous of them all, Halley's comet, the best pictures being delivered by ESA's Giotto probe. Since then no such opportunity had been available: although Giotto went on to visit comet Grigg-Skjellerup in 1992, its camera had been put out of action by dust impacts during the Halley encounter.

NASA's Deep Space 1 probe was launched in 1998, on a mission to test various new technologies, such as an automatic navigation system and lightweight components. It is also the first interplanetary spacecraft to use ion propulsion. Such engines deliver low thrusts, but may be operated for long periods, months or years at a time, whereas chemical rockets are fired only for seconds or minutes. Ion drives are much more efficient, and had already been used for station-keeping on Earth-orbiting satellites. This experiment proved that they could also be used for sending probes on longer voyages.

In 1999, Deep Space 1 flew past its first target, an asteroid called Braille. Some cynics have quipped that the name is apt, because the images returned were less than impressive, only a few pixels spread over the asteroid vaguely showing its shape. The encounter with comet Borrelly in September 2001 went much better. French astronomer Alphonse Borrelly discovered this comet in 1904. It laps the sun every seven years, but no ground-based photographs could match what Deep Space 1 saw as it passed within 1,400 miles of Braille on 22 September. The nucleus was found to be about five miles long, half of that wide, and shaped somewhat like one of the skittles used in ten-pin bowling. Erupting from that solid core were jets of dust and vapour, up to 60 miles long. These then disperse to form the vast cloud surrounding and obscuring the nucleus when studied from afar, and eventually the tail of the comet.

As with Eros, craters were found on the surface of the comet. It is a matter of conjecture whether these are the result of impacts by smaller projectiles, or cavities left after volatile ices vaporized from active vents. The mottled appearance of the comet indicates that it is not homogeneous. Comets vary widely in their properties, but it seems they are generally composed of a mixture of rock, water ice, and a variety of organic chemicals, although in differing proportions. The data obtained at comet Borrelly are still being sifted, but there is no doubt we will learn a lot from this close encounter.

The year 2001 may have seen major advances in our understanding of asteroids and comets, but a whole new era is dawning. Over the next decade a fleet of space probes will be sent to visit similar, but distinct, targets. NASA's Stardust probe is en route to comet Wild 2, charged with returning a sample of its dust to the Earth in 2006. Before then, another NASA mission, the Comet Nucleus Tour (or CONTOUR), should have returned close-up data from two comets and be on its way to a third. In July 2005, the Deep Impact probe is due to make its own crater in a comet, Tempel 1, telling us what lies beneath the surface. In the same year a Japanese satellite, Muses-C, should arrive at a near-Earth asteroid and drop a small rover on to its surface. And in 2003, the most ambitious project of all, ESA's Rosetta spacecraft, is due for launch. This is scheduled to fly past several asteroids before arriving at comet Wirtanen in 2011, and manoeuvre alongside as it comes closer to the Sun, monitoring the way the comet behaves as the solar heating increases. For those of us who study asteroids and comets, 2001 was a pivotal year. We saw things that we had never had a chance to see before, and we knew that a decade had begun that will revolutionize our knowledge of small solar system bodies, the astronomy closest to home.

During 2001, we got our best view of the heart of a comet. It had been a long wait

Invisible death

Alastair Hay

For two decades, scientists have been warning of the hazards of bioweapons research. Anthrax has become a terrorist weapon, but there is worse on the way

The US was better rehearsed than most for a terrorist incident involving biological weapons, but in common with most other countries, it expected the threat to come from the air. Scenario after scenario estimated casualties downwind of a bacteria-rich aerosol delivered by a relatively low-flying aircraft. The possibility of terrorists lacing envelopes with anthrax and using the postal service as the means of delivery was never considered. The route could not have been simpler. All it took was several envelopes in the post and the United States was reeling once again. This latest incident of bioterrorism in the US has left 5 dead, 18 infected with either the inhalational or cutaneous form of the disease, and thousands more on prophylactic antibiotics.

An earlier bioterrorist incident took place in 1984, when followers of the Bagwhan Shree Rajneesh treated food in a number of salad bars in the state of Oregon with *Salmonella typhmurium*, an organism commonly responsible for food poisoning. The Rajneeshees were in dispute with residents of a local town over a number of planning issues. Their actions led to the poisoning of 751 people; fortunately, none of the cases was fatal.

The attacks in the US were not the first in which biological agents were used deliberately to cause disease, nor are they likely to be the last. Pollution of drinking water supplies with the corpses of dead animals has been recorded in warfare since Grecian times, and was practised in both the American Civil War and the Boer War. In many conflicts, the bodies of the dead were far from sacrosanct, and the corpses of plague victims were thrown over the walls of cities under siege. This approach was reported to have been successfully employed by the Tartars in their war against the Genoese in 1346, and in the Russo-Swedish wars of 1710.

More underhand was the plea of the British Commander-in-Chief in North America, Sir

Jeffrey Amherst, to encourage the spread of smallpox among 'disaffected tribes of Indians'. As if anticipating this request, a subordinate had already acted, handing over, as presents to two Native American Chiefs, two blankets and a handkerchief from a smallpox hospital. The outcome is not recorded but can be imagined. Smallpox spreads by contact with contaminated fluids from the respiratory tract of the infected or from beneath scabs on the skin. In unvaccinated cases, 20–40 per cent of those infected die from the disease.

On the battlefield, bacteria and viruses are likely to be spread in a more purposeful fashion and in much greater quantities. In the short term, however, biological warfare has little effect – for the simple reason that bullets kill faster than bacteria. Its use, therefore, is more likely to be strategic. The delay between exposure and infection may allow the perpetrator to escape unnoticed, because the effects of his or her deed would not become immediately apparent. Where munitions are used, the requirement is that the organism should remain live and viable after dissemination from a grenade, shell, bomb or missile, or from the spray tank of an aircraft drone. These are but a few of the devices investigated for their usefulness, with some more effective than others, their suitability depending on both the agent spread and the ambient conditions.

The viability of organisms both before and after they have been turned into weapons is equally important. Many biological agents (anthrax excepted) have a limited life, their activity continually declining in storage unless steps are taken to slow the process down. Low-temperature storage or freeze-drying will help retain infectivity. Many bacteria and viruses remain viable after long periods of storage at minus 70°C.

Delivery of bacteria or viruses is most effective in a finely dispersed aerosol with particle sizes ranging from 1–5 microns – a size small enough to enable organisms to penetrate deep into the lungs. Agents that are unstable in liquid aerosols might be spread in a powder or slurry, the efficiency of dispersal being a key requirement.

The attacks in the US were not the first in which biological agents were used deliberately to cause disease, nor are they likely to be the last

When it comes to biological warfare, few agents can match the pedigree of anthrax

Ideal biological warfare agents might be those that resist environmental degradation as a result of temperature changes, humidity or ultra-violet light. Even anthrax, which is renowned for its environmental persistence, will not survive under all circumstances. An inappropriate temperature, rate of temperature change, degree of humidity or percentage of oxygen in the environment will prevent anthrax bacteria forming spores, the spores being a form of hibernation in which the bacteria can not only survive more extreme environmental changes, but retain their infectivity. It is the spores that are the means by which anthrax is transmitted in the environment and in warfare.

When it comes to biological warfare, few agents can match the pedigree of anthrax. Considered for use against Germany during the Second World War, preparations were well under way in both the United Kingdom and United States for a substantial programme to prepare anthrax munitions. Difficulties in production in the US prevented early delivery of the munitions and the end of the war in May 1945 precluded their use, originally envisaged to be early in 1946. Experimental programmes on the island of Gruinard, off the west coast of Scotland in the early 1940s, established the infectivity of anthrax when delivered from exploding munitions. Laboratory work in the British Government's chemical and biological warfare research establishment at Porton Down refined the calculations, and the US was employed as the manufacturing partner for the munitions, given that it was far removed from the risk of being bombed and the dangers consequent on anthrax being released into the air in huge quantities.

In parallel with this work in Europe and the US, the effects of the bacteria were studied on Russian and Chinese detainees, following their deliberate infection by Japanese scientists. At least 3,000 human subjects are believed to have been killed as a result of Japanese experimentation with a range of bacteria and viruses at Camp 731 in Manchuria between 1938 and 1945. Information on the nature of the infection caused by the test agents was certainly obtained in this experimental programme, but the bombs developed to transmit disease organisms were crude.

Two immediate beneficiaries of this Japanese research were the Soviet Union and the US. The former placed captured Japanese scientists on trial in 1949 for war crimes, an action dismissed by the West at the time as a mere propaganda exercise. The US, however, chose a different route. Japanese scientists involved in the Manchurian programme received immunity from prosecution for war crimes in exchange for research data. At the time of this exchange, the US had embarked on a significant offensive biological weapons programme. By 1969, having researched many candidate organisms, the US had at least eight standardized and ready for deployment in munitions. Ensuring a wide coverage, the US arsenal contained both lethal and incapacitating human pathogens and vastly greater quantities of two fungi that cause disease in cereal crops and rice respectively.

A unilateral order by the then president, Richard Nixon, led to the destruction of the US stockpile between 1970 and 1971, paving the way for a second international treaty outlawing the use of biological weapons in warfare. The 1972 Biological Weapons Convention (BWC), prohibits the development, production, stockpiling, transfer, acquisition or retention of weapons based on bacteria viruses or toxins. Defining how organisms that cause disease can be used, and in what quantities, the BWC, like its precursor, the 1925 Geneva Protocol, can only exhort signatories to respect its provisions; no international mechanism is available to police the regime.

Without an overseer, it is most unlikely that promises of good behaviour will be sufficient. Take the case of the former Soviet Union, a signatory to both the Geneva Protocol and the BWC. Obliged by the terms of both treaties not to engage in offensive biological warfare preparation, the Soviet Union appears to have been doing quite the reverse. Evidence from scientists defecting to the West indicated that a significant offensive biological warfare programme was being run by the state security service, the KGB. This programme continued until 1992, when it was stopped by the Russian Federation President, Boris Yeltsin, shortly after the programme became public knowledge.

Events prior to Yeltsin's intervention suggested the possibility of some preparation for biological warfare. The accidental venting into the air of a small quantity of anthrax from a military microbiology laboratory, in Sverdlovsk in April 1979, led to at least 64 deaths. Claimed by the Soviet Union to be an outbreak of intestinal anthrax, a subsequent joint US-Soviet investigation in 1992 confirmed that the deaths were from the inhalational form of the disease and that its victims were residents living downwind of the release on the day it occurred. The quantity of anthrax released can only be estimated, and it is not possible to say whether it put the Soviet Union in breach of its treaty obligations.

Defecting Russian scientists have made further allegations about the Russian programme. It has been claimed that at the Institute of Especially

Pure Biopreparations in St Petersburg, attempts were made to increase the infectivity of plague for use as a biological warfare agent. The plague bacterium said to be under research at the vaccine production facility was one that was already resistant to at least 16 different antibiotics. It was claimed that the programme was one to perfect the plague bacterium for use as an agent that would cause the pneumonic form of the disease. In this state, the bacterium is infective and disease spreads in contaminated aerosols when the victim coughs or sneezes.

Manipulation of biological agents is now routine in many research laboratories, and constitutes a legitimate approach to understanding the function and importance of the component parts of an organism. This research, however, might lead to organisms engineered for the deliberate spread of disease.

So far, most of the evidence suggests that when one performance characteristic of an organism is improved, another decreases. For example, the infectivity of a bacterium can be increased, but at the price of reducing its capacity to survive in the environment. For this reason, the prevailing wisdom is that the risk from biological warfare in the immediate future is not from manipulated organisms, but from existing agents. Recent work in Australia, however, in which the mousepox virus (closely related to smallpox) was inadvertently made much more virulent and lethal by

Evidence from scientists defecting to the West indicated that a significant offensive biological warfare programme was being run by the state security service, the KGB

incorporation of the gene for a well-known immunological messenger into its DNA, indicates why there is concern. The mousepox virus may be unique in the way it responds, but this may have implications for those researching viruses that cause human diseases.

Improvements, such as microencapsulation, in the technology of delivering medicines through inhalation may also open up possibilities for biological warfare enthusiasts. Encapsulation of vulnerable but dangerous viruses, or biological regulatory peptides for example, would provide the necessary protection for the agents. Delivery as aerosols would be easy. The consequences can only be imagined; the scenarios are infinite.

Defences against biological warfare agents will always be limited. The primary objective is preventing contact with the agent. Predicting which ones will be used in war or a terrorist attack is more guesswork, and hence vaccination, even if it were possible for a few, can never provide complete protection. Many vaccines confer only temporary protection. Early detection of any attack is a priority. Devices for screening pathogens in the air are at present available only to the military, but more widespread civilian application of this technology is likely and desirable. Good disease surveillance cannot be over-emphasized. Prompt recognition of an infection and of its type will probably remain one of the earliest indicators of any attack. Appropriately trained clinicians and laboratory personnel are key in the prompt diagnosis of a disease outbreak. Effective treatment options and appropriate public health measures will reduce the overall number of fatalities and casualties.

All of these measures to mitigate the effect of a disease outbreak can and should only be built into an existing public health programme. A 'stand-alone' system designed only to deal with an offensive biological warfare attack would be a waste of resources. Personnel in such an organization within a country might never be required to deal with a disease outbreak. It is far better to rely on a medical profession already trained and able to deal with the mass casualties likely to occur through any number of disasters.

Countries will continue to prepare for the possibility of an attack with biological agents. These preparations can only limit its effectiveness. In the longer term, prevention has to be the better option. Attempts to acquire a new international treaty with sufficient policing powers to control the deliberate use of biological agents in warfare are continuing. Such a treaty would inevitably require inspection of culture facilities, such as those employed by the pharmaceutical industry, which are capable of researching and producing the quantities of agents needed for biological weapons.

Unlike the chemical industry, which has agreed to intrusive inspections as part of an international regime to control the spread of chemical weapons, the pharmaceutical and biotech industries, particularly in the US, are reluctant to be part of an inspection regime on biological warfare. Unless they sign up to the idea of inspections, a robust treaty to prevent biological warfare – on a military scale at least – is unlikely. In the absence of a treaty, defence budgets in this area will continue to grow, but will never be sufficient to guarantee protection.

Making a beeline

Tim Radford

Bees had the animal behaviourists buzzing. Worker bees showed not only that they can work, but that they can also work it out. Scientists in Indiana reported in *Nature* in May that they could show that bees used 'optic flow' – the speed at which the scenery seems to rush past – to calculate the distance from hive to nectar. Bees need to know this because returning foragers famously pass on information about rich fields of flowers to their colleagues back in the hive. They do this by performing a tiny dance, the length of which is linked to the distance to food. The scientists trained the little creatures to fly down a featureless tunnel before heading for a feeder. Train travellers tearing across a featureless landscape notoriously feel that they are moving more slowly than when racing through forests. The bees flew a total of 11 metres, but performed a dance seven times longer than predicted for such a short distance. That sent their colleagues flying 72 metres in search of their supper, proof that the apparent speed of the passing landscape was the phenomenon that served as an apian odometer.

But *Apis mellifera* showed that it couldn't always be fooled so easily. In April, a team from Berlin in Germany, Narbonne in France and Canberra in Australia reported in *Nature* that they had trained bees to head down a plastic tunnel, read a signpost and choose the fork that would lead to the sugary reward. They did this by labelling the entrance to the tunnel with a colour, and then sticking the same colour on the turning that branched to the food. It seemed to them that bees that passed such a test and continued to pass it, might 'interpolate visual information, exhibit associative recall, categorize visual information and learn contextual information'. So they set up a new challenge. Instead of labelling the tunnel and its food-turning yellow, or blue, they suddenly presented the bees with black and white grating patterns, in radial, circular and linear shapes. They also swapped colours for odours – a choice of lemon and mango – and marked the maze accordingly.

The results were the same. The bees worked out what was going on and headed straight for the sugar. The scientists concluded that the bees carried in their tiny heads the abstract ideas of 'sameness' and 'difference' and applied them in reasoning. 'They can transfer the matching or non-matching ability to new stimuli for which the rules are not specified in the training,' they reported. In other words, they can think.

Our thermonuclear neighbour

Andrew Coates

When the sun sneezes, Earth catches a cold

The sun is a tormented ball of material in the plasma state, the fourth state of matter beyond solid, liquid and gas. The temperature is 15 million degrees at its core and 5,800 kelvin at the visible surface. Strangely, the atmosphere above is much hotter than this, up to a million or more degrees in the corona that is seen in an eclipse. We still do not fully understood why. Magnetic disruptions in the sun's hot atmosphere, less dense than any vacuum we can create on Earth, can have significant effects on mankind. During this solar maximum period in the star's 11-year cycle, many energetic solar eruptions have been seen, including – in April 2001 – the biggest measured solar flare of all time. Now, intense research on the sun itself and on Earth's magnetic defences is beginning to raise the possibility of predicting some of the effects of these huge events. We are partners in the dance: life depends on the sun's benign light and life is protected from the sun's harsher discharges by Earth's atmosphere and magnetic shield. This shield, which has its own lines of force, exists because the core of the earth is a huge fluid dynamo. On Mars, the dynamo stopped over 4 billion years ago, another reason why only early, simple life on Mars is likely.

Year of the record solar flare

2 April 2001. The sun has been very active in the last week. A huge amount of material from the corona, the sun's million-degree outer atmosphere, was suddenly flung from the sun on 29 March, as magnetic fields reconfigured. This time the material was fired towards Earth. It crossed the 150 million kilometres to Earth in two days. The 10-billion-tonne coronal mass ejection (CME) ploughed through the relatively calm solar wind between the sun and Earth on its way, causing a shock ahead like a sonic boom. Conditions were such that when it arrived on 31

March, this CME affected Earth's magnetic environment severely. Auroras, the northern and southern lights, were seen as far south as Mexico and Nice.

But today, 2 April, the sun is going to try a different trick. Its writhing magnetic fields are forming a new twist. Although we are watching the active regions on the sun using the solar satellites SOHO, Yohkoh and TRACE, we have no idea of what is about to happen, because the sun is too complex to predict at present. Suddenly, near the sun's edge, a cataclysmic explosion happens in the sun's close atmosphere. It has the power of a billion megatonnes of TNT. This solar flare unleashes huge amounts of energy in the form of X-rays, gamma rays and radio waves, which travel at the speed of light and are seen on Earth eight minutes later. The intense fluxes cause computer memories in some satellites to change state, and the radio noise affects radio communication, as with all solar flares. But this is no ordinary solar flare. Later analysis shows that it is the biggest on record.

This flare is not only the most powerful of all time: it is also one of a few which unleashes fast protons in all directions. It is not possible to predict which flares do this, although people are trying. The protons travel at a large fraction of the speed of light, reaching Earth some 20 minutes to an hour later. The subsequent radiation storm from the sun lasts two or more days. This too causes problems for satellite computer memories and is lethal to humans. No space walks take place during these proton storms and problems would be caused for astronauts outside Earth's protective magnetic shield.

The 2 April proton storm was not as severe as events in September and November 2001, or in July and November 2000, despite the fact that the flare itself was the biggest ever measured. We are lucky that the event happened near the sun's limb and that we are not in the direct firing line. Next time we may not be so lucky. Nevertheless, this event also launched a coronal mass ejection, which again moved away into the solar system, and not directly towards Earth. But satellites in Earth's orbit still saw the interplanetary shock two days later, as Earth was dealt a glancing blow.

Solar wind

As well as heat and light, the sun emits a constant but gusty stream of 'solar wind'. This is an electrically neutral stream of plasma, ions and electrons, moving away from the sun due to the heat of the corona. It is an excellent conductor of electricity, so it drags the solar magnetic field with it as it moves into the solar system. The solar wind is disrupted by impulsive events like CMEs, and wind that moves away from the solar poles tends to be faster, about 800–1000km/s, than that from more equatorial regions, which moves at about 400km/s. Also, the sun's 27-day rotation means that Earth is bathed in periods of fast and slow solar wind as the heliospheric current sheet rotates like a ballerina's skirt.

All this means that the solar wind conditions ahead of Earth's protective magnetic field can change dramatically. On average, the density of the solar wind there is about 10 particles per cubic centimetre – a billion billion times less than the density of air at sea level. The speed varies between about 300 and 800km/s and the temperature is on average 100,000K. Although the density is low, the equivalent electrical power incident on our magnetic shield is about 3 million megawatts – equivalent to mankind's current energy consumption. The amount of material hitting the shield is only about 40kg per second. The magnetic field strength is about 1/5000th of Earth's field strength at the equator. The north–south orientation of the field turns out to be vital in understanding what happens next. The field is as likely to be directed south as north.

Magnetosphere

If there were no solar wind, Earth would be like a bar magnet in space. The north pole of the magnet is currently in the southern hemisphere, so away from Earth the field is directed northwards. The charged particles of the solar wind are deflected by Earth's magnetic field. The Earth's field is compressed on the day side by the solar wind and pulled out into an invisible tail on the night side like a windsock. But these fluid concepts are not enough to understand the interaction. As the solar wind approaches Earth, it drags its solar magnetic field with it. If this field has a southward component, opposite to Earth's northward field, a key small-scale effect called magnetic reconnection can occur. As first proposed in 1961, and now spectacularly confirmed by the Cluster mission of satellites measuring the solar wind, oppositely directed fields that are forced together can suddenly reconfigure and reconnect. Near Earth, solar magnetic field lines connect through Earth and out the other side, becoming solar again at the other end. This allows the solar wind to flow past Earth, as the field peels across our magnetic shield like peeling a banana. The process itself is explosive, propelling plasma fast along the magnetic field towards the atmosphere, where it causes auroras.

As the field peels over the magnetopause, it eventually winds up in the tail – again oppositely directed, separated by a current sheet. Reconnection can occur again in the tail, shooting plasma towards and away from Earth. Exactly what triggers reconnection in the tail, and whether processes closer to Earth cause the trigger, are still not known. The net effect, however, is that plasma can be injected into the magnetosphere and can follow Earth's field towards the atmosphere, causing aurora on the night side of Earth.

Scientists do not really understand the reconnection process and what causes it, but some progress is being made. In the last year, evidence has grown that waves that are faster than were originally thought may be involved in the small-scale details of the process. However it works, it is a process that is ubiquitous in space physics, and probably plays the key role in solar flares and in coronal mass ejections, as well as several places where magnetic fields come together in astrophysics. Earth's magnetosphere is the best chance for understanding the process, as we can fly spacecraft through it.

The magnetosphere is a dynamic place. It contracts and expands as the solar wind changes. Sometimes the magnetopause (the 'shield' where reconnection on the day side occurs) can go inside the geostationary orbit at 6.6 Earth radii, where many commercial satellites are in orbit. The solar wind speed is important, too, causing flapping of the magnetosphere like waves across water. Both these parameters seem to control the acceleration of particles to relativistic energies inside the magnetosphere, feeding Earth's radiation belts where spacecraft fly.

What is space weather?

Space weather refers to conditions on the sun and in the solar wind, magnetosphere, ionosphere and thermosphere that can influence space-borne and ground-based technological systems and endanger human life or health. Our magnetic shield is not perfect. There are several timescales that are important for potentially dangerous space weather conditions. First, X-rays, gamma rays and radio waves from solar flares travel towards us at the speed of light, reaching Earth in about eight minutes. The X-rays can cause single-event upsets – memory-state changes – in satellite electronics, and enhanced ionization in the ionosphere, affecting communications, which can also be directly disrupted by the radio noise. Second, when energetic protons from flares are produced, they travel at a significant fraction of the speed of light, reaching us in about 20 minutes to an hour. Again, these can cause single-event upsets but are also extremely dangerous for astronauts, and can lead to an increase in secondary cosmic rays as they shower through the atmosphere, producing increased radiation levels at aircraft altitudes.

The third timescale is associated with coronal mass ejections. Depending on the speed at which these leave the sun and plough through the existing solar wind structure, they can reach Earth in 36 to 72 hours, with dramatic effects on the Earth's magnetosphere. At best, the magnetic shield is compressed, sometimes to within the geostationary orbit. This can disrupt orientation

> Suddenly, a cataclysmic explosion happens in the sun's close atmosphere. It has the power of a billion megatonnes of TNT. This is no ordinary solar flare... It is the biggest on record

systems that depend on Earth's magnetic field as a reference. At worst, if the interplanetary field is southward, enhanced reconnection results. This causes injection of solar particles through our magnetic defences on the day side, in the cusps and on the night side. These processes enhance the ring current around Earth, which can induce electrical currents in ground-based power and petrochemical distribution systems, and cause changes of up to 10 degrees in the magnetic field direction, as measured on Earth's surface.

If conditions are right, the injected particles can act as 'seed' particles for acceleration to relativistic energies, producing 'killer' electrons that can be dangerous to satellites. These particles can penetrate into coaxial cables and electronic components on satellites, depositing a charge that eventually builds up and discharges, causing a spark. This process is called deep dielectric charging. Satellites can also be affected because they are immersed in a sea of charged particles. This can cause charging on non-conducting parts of a satellite's surface ('surface charging').

Usually, auroras are confined to ovals surrounding Earth's magnetic poles, where magnetospheric particles collide with and excite gas molecules. The ovals usually occur over northern Scandinavia, Russia, Canada and Antarctica. At times of enhanced activity, the ovals can move further away from the poles. The atmosphere can also be heated by the increased currents flowing in the ionosphere at disturbed times, causing the atmosphere to expand and increase drag on satellites in low Earth orbit (400–1,000km above the surface).

Energetic particles from sources beyond our solar system are strongly affected by the magnetic field patterns in interplanetary space and by Earth's field. Nearer the magnetic poles, it is easier for these energetic particles to reach the top of the atmosphere, again causing secondary cosmic ray showers. Naturally, these are more intense near the poles than towards the equator, and this has implications for commercial aircraft routes.

Examples of effects on space and ground-based technology

Satellites have been lost, or operations disrupted, by space weather events. In January 1997, Telstar 401, a communications satellite, was lost, most probably as a result of deep dielectric charging by 'killer' electrons. This event was unusual in that the solar cause was seen and the effect predicted. The same process is suspected in the loss of Galaxy IV in 1998 and two Canadian Anik satellites in 1994. In 2000, the large solar event in July, called the Bastille Day event, caused the loss of control of a Japanese X-ray astronomy satellite, ASCA, through atmospheric expansion. During this same event, the efficiency of solar panels on all satellites, including two of the newly launched Cluster satellites, was reduced by radiation damage. As commercial communications satellites, such as Telstar and Galaxy, can cost some $250million to replace, this is clearly a major problem.

13 March 1989, 2.45am For no apparent reason, power distribution systems in Quebec suddenly tripped out. Quebec had no electrical power for some 9 hours, affecting 6 million people. Later analysis showed that this had been caused by a large solar event and its subsequent effects in the magnetosphere. There have since been other examples of events caused by solar activity, from brief interruptions to transformers burning out, all caused by geomagnetically induced currents (GICs) from huge currents in the magnetosphere.

The science behind space weather

The fleet of satellites studying solar-terrestrial physics is now larger than ever. Since 1996, we have been able to watch solar events from cradle (the sun) to grave (the magnetosphere and ionosphere). Missions include: SOHO, Yohkoh, TRACE studying the sun; ACE and SOHO measuring solar wind conditions upstream of Earth; Polar, Wind, Geotail, Interball, IMAGE and now Cluster studying the magnetosphere; and FAST studying the ionosphere. In their own ways, all of these missions are contributing substantially to our knowledge of the solar-terrestrial system.

With this armada and an army of scientists studying the data, how good are we now at predicting space weather events? Let us examine the various effects:

- **Solar flares**. There is currently no way of reliably predicting flare events or flares that produce protons. Work is under way to examine magnetic structures in the sun's atmosphere for this, but current techniques are unreliable.
- **Coronal mass ejections** – prediction. Using Yohkoh images, scientists have discovered S-shaped 'sigmoid' features in the million-degree solar corona, which seem to precede some coronal mass ejections. While the technique is not reliable,

examining solar magnetic field complexity may provide the key. Reliable prediction schemes may have to await 3-D magnetograph data from the Solar-B mission in 2005.

- **Coronal mass ejections** – propagation. Coronagraph and artificial eclipse data from the SOHO-LASCO satellite are proving extremely important in identifying coronal mass ejections and tracking them near the sun. Billowing material in our direction, events headed towards Earth appear as 'halo' events. Continuous coronagraph data is vital for identifying these and for measuring their launch speed from the sun. However, one cannot tell from using a single satellite whether the event is in fact headed away from us or slightly to the side, missing Earth altogether. The Stereo spacecraft will provide a three-dimensional stereo view following launch in 2006, but even then, it will not be possible to calculate the time at which such an event will reach Earth with any certainty.
- **Solar wind changes** – effects on the magnetosphere. Despite many years of work and many satellite missions, we still do not know exactly what triggers reconnection and magnetic storms, and thus the acceleration of particles to relativistic energies and enhancement of the ring current, although we have some promising clues. Research work using several satellites, particularly the four-spacecraft Cluster mission, is aimed at understanding the trigger for reconnection and storms, and the acceleration process for the damaging relativistic ('killer') electrons. The Magnetospheric Multiscale mission (launch 2006) will be important here too. In the more distant future, a multi-spacecraft mission like SWARM, with 30–50 satellites, may be able to track space weather fronts as they traverse the magnetosphere.

Applied research

Our group at UCL's Mullard Space Science Laboratory is pursuing two contracts funded by the UK satellite insurance industry, one to produce a prototype 'black box' for commercial satellites to measure the number of killer electrons, and the other to produce 'space weather forecasts' using measured solar wind conditions. UCL-MSSL is also working with civil aviation companies to monitor radiation in aircraft cabins, and similar studies are in progress across Europe. The British Geological Survey is working with power companies and oil prospecting companies to assess and predict space weather effects on those industries. JHU-APL is using data from magnetometers on the Iridium fleet of satellites to develop magnetic field models that can predict Earth's response to the solar wind onslaught. And several groups in the UK are working on parallel ESA contracts to define a future European space weather programme.

Eclipse over Westminster

Many sections of industry are deeply concerned about space weather effects. This interest will grow as satellite technology becomes more integrated. Although flares and coronal mass ejections peak at solar maximum, they still occur at other times within the activity cycle. For CMEs, for example, the rate changes from an average of four a day at solar maximum to perhaps a quarter of this at minimum. Solar wind-driven magnetic storms lag behind the solar maximum by about 1–2 years. And information on satellite problems and failures shows that losses can occur throughout the solar cycle.

The future

The current state of space weather prediction is comparable to that of terrestrial weather prediction 50 years ago. We need to understand, or understand better:

- how and why continuous and impulsive emissions from the sun occur
- how they propagate between sun and Earth, and how our magnetic shield is penetrated
- the effects of solar wind changes on Earth's magnetosphere and radiation belts, where the satellites used in our daily lives are stationed and where 'killer' electrons are found
- and finally, the coupling of electromagnetic energy all the way through from the sun to our atmosphere.

The 'Living with a Star' research programme is designed to improve our understanding in these areas. It will involve the international community and a range of missions to study the sun-Earth connection. And it will play an important role in the science behind space weather over the next 10 years.

An image in your head

Joe Herbert

Neuroscientists can now literally read your mind as you wrestle with unfamiliar faces, complicated syntax and the sound of music

The living brain doesn't look as if it's doing anything. Hearts beat, muscles contract, guts squirm, but the brain just seems to sit there. About 100 years ago, however, Hans Berger discovered, using what became known as the EEG, that the brain emits waves of electrical activity that alter as you think hard, or sleep. Was this the long-sought window into the workings of the brain? It hasn't turned out that way: while the EEG has been useful for studying sleep, and for defining certain types of epilepsy, it has been disappointing as a tool for understanding what happens in the living human brain during such interesting activities as movement, sensation, thinking, emotion and speech.

Recently things have changed. New methods of imaging the brain have rekindled scientific interest in these questions. There are a number of different techniques, but MRI (magnetic resonance imaging) is particularly interesting. MRI has the advantage of being non-invasive: that is, unlike some other methods of brain imaging, the subject doesn't need to be exposed to X-rays or other forms of radioactivity. It's totally harmless, as far as we know, although an MRI machine is expensive (£1–2million). MRI enables one to see a series of slices through the brain of living subjects, allowing the size and shape of individual bits of the brain to be measured, and problems such as strokes and tumours to be detected. It's been a boon to clinical neurology, letting doctors see and diagnose abnormalities in the brain with a precision only dreamed of just 10 years ago. But that's not all. MRI can also detect changes in the amount of oxygen carried by the blood. The brain has exquisite control over its own blood supply: if one part of it works harder than another, then it uses more oxygen, and so the blood supply of that bit of brain alters. This functional change can be detected by MRI, which is now called fMRI (functional magnetic resonance imaging). Now we have two pieces of information: an

> Faces that are racially different from the observer have a greater effect on the amygdala than same-race pictures: a neural basis for racial prejudice, perhaps?

accurate picture of the living brain, and a measure of the amount of oxygen each area is using (its activity) at the moment the picture was taken.

fMRI can thus be used to study patterns of local activation in the brain. For example, what happens in the brain when we think about something? Or feel an emotion? Or listen to music? Or understand a spoken or written sentence? Or compose our reply? Or learn something? Only recently has there been real progress in answering such questions. Using fMRI requires very careful and well thought-out tests (that is, good neuropsychology), a high level of technical ability (fMRIs that have good resolution so that one can see different bits of the brain clearly) and complex statistical methods. Unfortunately, people who know about psychology don't often know much about the physics of scanning; most physicists are bemused by psychologists, neuroscientists are not always good at psychology (and vice versa), and the statistics is tough for all of them. But in the last year or so these disciplines have begun to pull together, with some striking results.

The cerebral cortex is the great thick sheet of neural tissue that surrounds the rest of the brain, crumpled into the skull and giving the brain its characteristic appearance. We know that the cortex carries out the most elaborate neural functions, and that different parts of it do different things: for example, one cortical area is concerned with generating movement, another with vision, a third with hearing and so on. Other areas of the cortex do even more complex things: decision-making, memory, language, social interactions and so on. We know this because damage to different parts of the cortex results in the loss of different abilities.

Nineteenth-century clinicians, using brilliant insights from brain-damaged patients, recognized two main cortical areas (usually on the left side) important for language: one that generated speech, and another behind it that recognized spoken language. What we didn't know, until recently, is what these areas do during normal activity in people who are healthy. Cognitive psychologists, however, have long suspected that the neural machinery processing language is more complicated than this. Recent fMRI studies support this view. Words that are read activate a different cortical area from words that are heard; other areas respond to the speaker's own voice, or are activated by recalling words, or process sentences (syntax and grammar) rather than individual words (lexicon). Areas of the brain that perform defined functions are called modules. There may be a large number of cortical modules, each specializing in particular aspects of language comprehension and production. This has very important theoretical implications: how does each module recognize its particular language component? How do the different modules work together to form the seamless language abilities familiar to us? What happens when we learn our original language or a new one? fMRI shows that these modules are already functioning in quite young children (who are, of course, the most adept at learning language).

Most people love music of some sort. fMRI has shown that the areas of the cortex activated by music differs in skilled musicians (theirs is on the left side of the brain) and those who can't play (on the right). These are different areas from those responding to language, although listening to songs (as you might suspect) can activate 'language' as well as 'music' areas. Are potential musicians born with different brains, or is there a change in their brain architecture during musical training? We now know that the brain is much more malleable than was once thought, but this would be a striking example.

Emotional experience also activates particular parts of the brain. Just below the cortex, above your ear, is a round structure about the size of a nut – the amygdala (amygdala means almond). fMRI shows this structure to be activated by emotional, particularly fearful, experience or sights, supporting the idea that the amygdala is concerned with either the experience or recognition of emotion. Faces that are racially different from the observer have a greater effect on the amygdala than same-race pictures: a neural basis for racial prejudice, perhaps? Sexual stimuli also activate the amygdala, but if the subject consciously inhibits an erotic response then the amygdala no longer responds – a region of the cortex at the front of the brain (which connects to the amygdala) is activated instead. This may represent the neural machine controlling emotional responses. Not far from this frontal area is one that responds to rewards (e.g. being given money); it's also activated by jokes, which is perhaps why we like hearing them.

By now you may be pretty impressed with fMRI, and you would be even more impressed if you were to visit an fMRI laboratory. This is high tech of cinematic proportions. People crouch over computer screens, lights blink satisfyingly and the machine itself makes rather a noise. But think for a moment what all this technology is actually doing: measuring changes in blood flow (a kind of neural blush) in parts of the brain. Even if this faithfully reflects the activity of local brain cells, it tells us only about the volume of activity, nothing about its nature. How do we recognize the face of a friend? How do we learn to be socially adept? How do we decide to take the dog for a walk or read this book? Why does activation of the amygdala cause the emotion of fear? fMRI won't tell us how the brain works, simply which parts we should be looking at. To understand what actually goes on in the living brain is another story.

Ray of light

Tim Radford

Light sets the ultimate speed limit: 300,000km a second in a vacuum. Light slows as it darts through water, or through a diamond, but not by very much. Massachusetts scientists shook the world of physics when they reported in *Nature* in January that they had slowed a beam of light to a standstill, held it still for a thousandth of a second and then watched it spring away again at the speed of light.

The scientists shone a laser pulse through a strange form of matter called a Bose-Einstein condensate. This was a tiny cloud of sodium atoms cooled to a billionth of a degree above absolute zero (minus 273°C), trapped in a magnetic field inside a vacuum chamber. It was, during the experiment, the coldest place in the universe. At that temperature, all the atoms behave as one atom: a condensate, or a kind of atomic syrup. ('A cold atom cloud in optical molasses looks like a little bright sun, 5mm in diameter,' Lene Vestergaard Hau of the Rowland Institute for Science reported later in the year in *Scientific American*.)

The team's first triumph came in 1999, when they managed to make the light pulse dawdle at a speed at which it could have been overtaken by a bicycle. In 2001, they reported the ultimate: light at a dead stop. They fired a kilometre-long pulse of light at 300,000km a second. The leading edge went through the vacuum chamber window and started to hit the tiny cigar-shaped cloud of ultra-cold atoms. The front moved incredibly slowly, while the rest of the pulse caught up and began to concertina into something that would fit inside the sodium cloud. Then the researchers switched the laser beam off and on again. In a sense, the light ceased to exist, but its hologram remained stored in the gas cloud. For one thousandth of a second – almost the time it would take light to get from London to Birmingham and back – the entire kilometre pulse was held still inside a cloud of optical molasses the size of a fingertip. Then it sprang away again at 300,000km a second.

A second team at the Harvard Smithsonian Centre for Astrophysics duplicated the same experiment in a slightly different way. The effect was, said a commentator in *Nature*, as if a high-speed train ran into a sheet of gossamer fabric and vanished. 'The roar of the train is replaced by an eerie silence. Then, as suddenly as it disappeared, the train emerges from the other side of the fabric, resuming its original length and speed, and roars out of sight.'

In a sense, the light ceased to exist, but its hologram remained stored in the gas cloud

The man who could only read numbers

Brian Butterworth

Memory seems to be a house of many chambers

We were not very surprised that Mr Harvey (not his real name) could still calculate accurately, even though brain damage had deprived him of speech to such an extent that he could not name even the commonest objects. Three patients had already been reported whose mathematical ability had survived brain damage, but whose language had not. We were mildly surprised by his ability to read aloud the numerals 1, 2, 3, and even four-digit numbers such as 7,495 without error. Given his enormous difficulties with speech, this was unusual, but not completely unknown. Was it relevant that Mr Harvey, before his retirement, had been a banker and that his hobby was gambling? This would have given the mathematical muscles in his brain more exercise than most people's, and we knew that brains, like real muscles, atrophy without exercise. Practice rewires the brain, so it was likely that the mathematical part of his brain had recruited more cells and had strengthened the connections between them. So if Mr Harvey started with more tightly interconnected brain cells devoted to mathematics, a reasonable level of number skill might well have survived three years of an insidious disease eating away at his brain.

What astonished us, though, was his word reading. He was quite unable to read even the simplest and commonest words – 'for', 'tree', 'then', 'take', 'give', 'you' – but he had no trouble at all reading number words. Though he couldn't read 'you', he could read 'two', and 'three', 'ten', 'eight' and 'five', not to mention 'thousand' and 'million'. This was quite unprecedented in the annals of neuropsychology. What could be the explanation of such a weird anomaly, and could it tell us anything about the nature of normal reading, or indeed about normal mathematics?

Patients with these very specific and unusual patterns of preserved and defective abilities are like rare archaeological specimens. We can't go into a laboratory and make them. We just have to find them, and to do this, like the good archaeologist, we need to know what to look for. And, perhaps more importantly, we need to know what we've got when we find it. This takes both acute observation, and a wide knowledge of theory. 'Theory' in this case means having a view about how different mental abilities are organized in the brain. In practice, this translates into having an idea of what can go wrong by itself and what cannot.

These specific disturbances of cognitive function have been responsible for major advances in our understanding of the mind. We now know, for example, that amnesics – the Lost Weekend type of amnesics – do not forget everything. Even the most severe cases are still able to remember how to speak, to read, that Paris is the capital of France, that kangaroos are not indigenous to Southend-on-Sea, and to calculate. The part of memory that is responsible for this preserved

The part of memory that is responsible for this preserved information is called 'semantic memory'

information is called 'semantic memory'. Remembering who you are, what you did yesterday or last year – 'autobiographical memory' – is a function of a brain structure called the hippocampus, while semantic memory uses other brain regions. In Ken Follett's thriller *The Code to Zero*, the hero tries to recover his autobiographical memory by using his semantic memory. Finding that he understands the technicalities of a rocket launch tells him that he must have been a rocket scientist (I won't ruin your enjoyment of the ingenious plot by revealing more).

There is also 'procedural memory' for skills such as driving or skiing. A few years ago, a distinguished Italian neurologist and expert on memory disorders, Dr S, was skiing too fast, as usual, and took a tumble. When the rest of his party caught up with him, he stared at them strangely. He was surprised to find his wife and his best friend suddenly looking very old. And he failed to recognize some of his younger colleagues. The neurologists in the party did a piste-side examination, and diagnosed autobiographical memory loss, probably of the last 25 years or so. This is why his wife and friend looked old, because his memory of their faces was 25 years old, and of course, he had come to know his younger colleagues within the 25-year blank period. However, he was still able to ski back to the bottom station, where he declared: 'See, my procedural memory is still intact!'

He was taken to the local hospital. Here he promptly asked for a brain scan, a procedure that he would have learned about during the blank quarter of a century. A scan was carried out, and Dr S looked at the results and made his own diagnosis. Seeing his name and age on the scan, he noted: 'There, the brain looks in very good shape for my age; no signs of atrophy. I must have transient global amnesia.' This was accurate, and showed that the semantic memories created in this period were still accessible. Global amnesia meant that not only was he unable to remember the past, but he could not create memories of new events. Fifteen minutes later, he said: 'Shouldn't I be having a brain scan?' This condition was, thankfully, transient and after 20 hours, he recovered completely. My impression is that he now skis a little more carefully.

Mr Harvey is the exact opposite of Dr S. His autobiographical memories are reasonably well preserved. We asked him to read the word 'theatre'. He couldn't pronounce the word, but he did say: 'I delved into that every year in Oxford from 1950 to 1960.' His wife confirmed that indeed he had been a keen amateur actor and director during this time. His problem lay in his semantic memory. He had very few words left – 'delved' coming up very often in his speech. In a standard clinical test, he failed to name a single common object, like table, chair, clock, or glove. And he couldn't point to the right picture when you said these words. He was unable to classify pictures of animals into those living in England and those living outside England. This combination of good autobiographical memory and poor semantic memory caused by a neurodegenerative disease is called 'semantic dementia'.

For two years, Mr Harvey had been in the care of one of the world's leading experts on semantic dementia and other memory disorders, Professor Michael Kopelman of St Thomas's Hospital in London. Kopelman and I were interested in how numerical knowledge and calculation skills relate to other aspects of memory, and we had noticed that in reports of other semantic dementia patients, these abilities could be preserved even when much else had been lost. On our team, which is funded by the Wellcome Trust, there was an enthusiastic Italian neuropsychology graduate called Marinella Cappelletti, who was to do most of the testing of Mr Harvey, including going to his home an hour and a half away when we needed to collect more data. (When I hear Europhobes banging on about EU waste or petty regulations, I want to shout out 'But what about ERASMUS?' This is a programme, paid for by our contributions to the EU, which enables students to spend time at a university in another EU country. And it's a wonderful thing. Students get a chance to see what it's like to be a student and thereby to live and work in another country, to become less parochial. I have been a beneficiary of this. A stream of brilliant European students have worked in my lab over the years, and some, like Marinella, stay on to do PhDs and get jobs in our research institutes and universities. This is not only good for British science, but it helps to create a European science base that can compete with the US and Japan.)

We devised a battery of numerical tests and systematically tested Mr Harvey more or less once a month for over a year. Mr Harvey always arrived at the clinic immaculately dressed, usually in a pinstripe suit and tie, and was always smiling and affable, but it became apparent that his condition was getting worse. He was losing his vocabulary, and was performing more and more poorly on our tests of semantic memory. However, for most of this period, his calculation remained at a very high level. His arithmetic, both oral and written, was usually without error. He could do long multiplication flawlessly. He could read long numbers and write them to dictation without difficulty.

We had now demonstrated conclusively, for the first time, that numerical skills could be preserved when most of the rest of semantic memory, including words, were lost. This is important because it refutes the idea, popular with both psychologists and laymen, that adult calculation is carried out in language and that arithmetical facts are stored just as verbal formulae.

However, we still did not know why he was

able to read only number words, and as we soon discovered, these were also the only words he could write. It wasn't just that number words were the only words he could remember. After all, he chatted with us, albeit hesitantly, about things other than numbers. So, we asked him to read words that he had used correctly in conversations, to be sure he knew them. He was slightly better with these words than assorted common words; but since he was able to read number words 100 per cent correctly every time, this could be only a tiny part of the explanation. Memory research wasn't going to provide the answer.

We turned to research on reading. In 1973, two Oxford neuropsychologists, John Marshall and Freda Newcombe had published a study of three patients whose reading had been affected by brain damage. In this study, they had done something that had never been done before. Previously, reading disorders, called alexia, had been classified into two types: alexia with agraphia (writing disorders) and alexia without agraphia. Marshall and Newcombe took the revolutionary step of not just counting the words correctly written, but analysing the kinds of errors that the patients made when they attempted to read a word. They found that one of their patients, a skilled reader before his brain damage, frequently mispronounced one or two letters in the word rather in the way that a child who didn't recognize the word might. For example, he read 'insect' as 'insist', softening the 'c'; and he read 'listen' as 'Liston, the boxer', sounding out the 't', which curiously is silent.

Two other patients made characteristic errors that had never been noticed before, reading 'bush' as 'tree', and 'ill' as 'sick'. Obviously, they couldn't make these errors simply by sounding out the letters in a childish fashion. They must in some way have read the words correctly, and retrieved their meanings. Somehow, in the process of trying to say the words they had come up with a word that was similar in meaning, but not in sound, to the target. Pondering on this puzzle, Marshall and Newcombe came up with a theory that became the standard in reading research (though nowadays it has a few more bells and whistles). They argued that when we see a word, we use two 'reading routes' automatically and simultaneously: we try to sound it out letter by letter, a kind of mental phonics process, and at the same time we try to retrieve the meaning of the whole letter string without bothering about its pronunciation. The patient who said 'Liston, the boxer' was using letter-to-sound route only, presumably because the other route had been damaged. The other two patients were using only the reading via meaning route, again presumably because the other route had been damaged.

Of course, 'bush' doesn't mean the same as tree, so why did these errors arise? It seems that we

We read far faster than we normally hear words – 200 to 300 words per minute, 3 to 5 words per second

need both routes. We need to be able to recognize words as wholes, since many have irregular spellings, or pronunciations that depend on knowing their meaning in context – such as 'lead' and 'wind'. And the letter-to-sound route is needed in case we have never seen the word before. With two routes operating together, one can check the output of the other. This may be particularly important given that we read so fast, far faster than we normally hear words – 200 to 300 words per minute, 3 to 5 words per second. With one route dysfunctional, errors will arise, with the kind of error depending on which route is affected.

This idea gave us the crucial clue to Mr Harvey's reading. Suppose he wasn't able to use the letter-to-sound route at all – in that case he would have to rely exclusively on reading by meaning. We tested this by seeing if he could read letter strings he had never seen before, such as 'zind' and 'yead'. He couldn't. We also looked to see if he made 'regularization' errors, such as pronouncing the 't' in 'listen', reading 'pint' to rhyme with 'hint'. He didn't. Normal readers using both routes are more accurate when reading regularly spelled words than irregularly spelled ones because the two routes interact. Mr Harvey was no better at reading regular words because the letter-to-sound route wasn't working.

The words he could and could not read were defined solely by their meaning. If they referred to numbers (including ordinal number words – 'first', 'second', 'eighteenth'), he could read them. If not, he was unable to. This reinforced our view that he was reading exclusively by meaning. We knew from our other tests that his semantic dementia had severely affected even his knowledge of the meaning of common words for everyday objects. But it was clear that he understood the meaning of number words, since he was still able to carry out flawlessly tasks that depended on knowing these meanings. So he could say which of two numbers was larger, even up to four-digit numbers. And, of course, he could calculate. So it is not just that he knew that 'seven' denotes a number, he knew precisely which numerical value it has – not six and not eight, and that added to five, it makes twelve. In other categories of

I think many dyslexics would benefit from more emphasis on a whole-word, meaning-based, approach to reading, and less on phonics

knowledge, even when he wasn't completely at a loss, his grasp of concepts and meanings was much vaguer, as is usual in these cases. As his disease progressed, he found that he was unable to recall some of the facts he'd learned in school, such as his multiplication tables. But he compensated well enough by using successive addition to solve the problems we gave him. This was, in a way, even more impressive than being able to retrieve the product of, say, 8 x 7, which could be mere rote memory, as it showed that he still understood both number meanings and the concepts of arithmetic.

The solution to his extraordinary reading turned out to be straightforward. His disease had left him reading by just one of the two routes we normally use. He could read only via meaning. The only meanings clear in his brain were number meanings, so it was just these that he was able to read accurately.

Now Mr Harvey's case is not just a medical curiosity. Nor is it just an example of one man's struggle to overcome an unpredictable and crippling disease, or a pointer to how far neuroscience has progressed by 2001. It solves two problems – though it raises more questions, of course. First, we have confirmed what theorists had previously only speculated, that reading via meaning is sufficient for accurate pronunciation, since Mr Harvey can read number words with perfect accuracy.

This may give us a clue as to how to help dyslexic children. One of their main problems is in learning the sound of each letter. This is particularly hard in English, where letters can have very different sounds depending on the word they are in. Just think of 'g' in the following words: 'tug', 'tough', 'though'. A dyslexic young woman we studied some years ago suffered six years of trying to learn to read through phonics, and was almost classified as 'educationally subnormal'. However, when her mother sent her to a different school that taught reading by the whole word, look-and-say method this worked excellently for her. We found her reading to be efficient, but unusual. She read most words so well that on standard tests she would not have been classified as dyslexic. However, she was quite unable to read new words. She always had to ask someone to read them to her, and then she would try to remember how they sounded.

I think many dyslexics would benefit from more emphasis on a whole-word, meaning-based, approach to reading, and less on phonics. Our study of Mr Harvey shows that this single route can work effectively (though not as effectively as the two routes together, of course).

Many years ago, the great British neuropsychologist, Elizabeth Warrington, discovered that neurological patients could have selective impairments of single categories of knowledge within semantic memory – they may still know about living things and foods, but have lost information about furniture, for example. In these studies she did not specifically consider numerical knowledge. The case of Mr Harvey, along with converse cases where language is spared, but numerical abilities severely damaged, shows that numerical knowledge is separate in the brain from our knowledge of language and the rest of semantic memory. This is somewhat counter-intuitive, given that most of what we know about numbers was learnt through language, and we can all remember hours of tedium reciting our tables in a singsong voice. But however we learned about numbers, they end up in a region of the brain known as the parietal lobes. Knowledge of language is mostly in the dominant frontal lobe (Broca's area) and Wernicke's area in the temporal lobe, near where most of the rest of semantic memory is located. Brain scans showed that disease was ravaging Mr Harvey's temporal lobes, but had left the parietal lobes intact, showing that different categories of knowledge may be in quite separate lobes of the brain. Why number knowledge should be in the parietal lobes and not in the temporal lobes is a question we are currently pursuing.

Our study left us with one further puzzle. Mrs Harvey had told us that her husband, an inveterate gambler on the horses, now bet only on dogs. Why, we wondered, had semantic dementia driven him to the dogs? This is where my misspent youth came in useful. I remembered that one bets on horses by giving the horse's name: 'Ten pounds to win on Galileo, please.' If you can't read the name, this is going to be difficult. But since all dog races have six starters, it's normal to say 'Ten pounds to win on number six, please.' This Mr Harvey can do. I hope he has better luck with dogs.

The old ladies know best

Tim Radford

Old matriarchs have a vital role – at least in the African savannah. In the elephant world, when it comes to keeping the family lore – and a wary eye on strangers – grandma knows best.

Sarah Durant of the Institute of Zoology in London and Karen McComb of the University of Sussex studied 20 small family groups of females and their calves at the Amboseli elephant research project in Kenya. The groups move around, foraging for food independently, encountering other groups from time to time. The two researchers were interested in what happened as groups met. They played recordings of elephant calls and watched the responses of the group. The elephants behaved just as humans tend to do. When they heard calls from complete strangers, the mothers clustered around their young, protectively. When they heard the familiar noises of relatives or old friends, they carried on grazing without showing wariness or suspicion.

But some groups were a lot better than others at telling known friend from potential foe. The researchers tried to work out what made the difference. The number of calves, the number of females and the mean age of the females were all ruled out as factors. What did matter was the age of the oldest female. The matriarch or 'grandma' figure had a greater store of knowledge of group behaviour and family alliances, and was better able to tell which were old friends, and which were unknown quantities. The two scientists then matched seven years of data from these audio playbacks with 30 years of observation of around 1,700 elephants in the region. This showed that matriarchs paid their way, in evolutionary terms. Groups with the oldest females tended to produce the most young per female. Social experience was a powerful factor in survival.

'It all boiled down to what age the matriarch was. Families with older matriarchs were massively better at picking out genuine strangers. When faced with more familiar individuals, they would remain relaxed. But they pinpointed the ones that would have constituted a genuine threat,' Dr McComb said. 'We believe this to be the first statistical link between social knowledge and reproductive success in any species.' The catch is that the oldest females were likely to have the biggest tusks – and present the most tempting target for poachers. So the finding has powerful implications for conservationists too. What goes for elephants might be true of other social creatures. Female sperm whales also wander around in little communes with their young, occasionally meeting other sperm whale groups. Low birth rates in many places might be explained by the toll of whaling practised until 18 years ago. Hal Whitehead, a marine biologist of Dalhousie University in Halifax, Canada, told *Science* magazine of 20 April: 'When you poach an animal, you are not just taking one life away; you are taking away the influence of that animal on other animals.'

Picture credits

p. 25: Murdo MacLeod
p. 38: NASA
p. 49: David Sillitoe
p. 58: NASA
p. 68: David Sillitoe
p. 87: Kenneth Saunders
p. 103: Graham Turner
p. 111: Garry Weaser
p. 125: David Sillitoe